THE STARDUST REVOLUTION

THE STARDUST REVOLUTION

THE NEW STORY OF OUR ORIGIN IN THE STARS

JACOB BERKOWITZ

59 John Glenn Drive
Amherst, New York 14228–2119

Published 2012 by Prometheus Books

Cover image of the Tycho supernova remnant originally by
MPIA/NASA/Calar Alto Observatory and modified by Jacqueline Nasso Cooke
(unaltered image available in photo insert)
Cover design by Jacqueline Nasso Cooke

Inquiries should be addressed to
Prometheus Books
59 John Glenn Drive
Amherst, New York 14228–2119
VOICE: 716–691–0133
FAX: 716–691–0137
WWW.PROMETHEUSBOOKS.COM

16 15 14 13 12 5 4 3 2

Library of Congress Cataloging-in-Publication Data

Berkowitz, Jacob.
The stardust revolution : the new story of our origin in the stars / byJacob Berkowitz.
p. cm.
Includes bibliographical references and index.
ISBN 978–1–61614–549–1 (cloth : alk. paper)
ISBN 978–1–61614–550–7 (ebook)
1. Exobiology. 2. Life--Origin. 3. Stellar dynamics. 4. Cosmology. I. Title.

QH326.B47 2012
523.1--dc23

2012023387

Printed in the United States of America on acid-free paper

For my parents,
Morton and Birgitte Berkowitz,
with love

CONTENTS

PART 3. THE LIVING COSMOS

PROLOGUE

EXTREME GENEALOGY

I've always felt the tug of questions of identity, of who and what I am. These questions ring all the more loudly in middle age: I see parts of me emerging in my son's gangly adolescent limbs, and my father's septuagenarian facial traits are reflected in the mirror. I see a long line of connection, of which I am one bead along the string. For me, this long string of connection is a story, and this book is one telling of that story.

It didn't start out that way. I thought I was writing a book about the intersection of evolutionary biology and astronomy. But in the strangest way, this book is a form of shared autobiography.

The Stardust Revolution is the story of the greatest genealogical search of all time. It's extreme genealogy. It's extreme in terms of time, connecting us back to the very beginnings of time and space in the big bang. It's extreme in what it tells us about the nature of our ancestors. They were stars. We usually think of genealogy in terms of readily recognizable traits—"She has her grandmother's eyes," or "He's a beanpole, just like his dad." In extreme genealogy, the traits we're talking about aren't recognizable in family photos or with a glance in the mirror. It's not our hair color or the shape of our hands that's of interest. The elements are more fundamental: the types of atoms from which we're composed, the chemical bonds between them, the molecules that make up our cells. This is how our indelible stardust heritage is apparent in our blood and bones.

In pulling together the disparate threads that connect us to the past, genealogists follow various trails of evidence. There are the documents, including birth, marriage, and death certificates. There's the living memory, such as a grandmother's amazingly clear recollections of *her* grandmother from the previous century. There are the tidbits of other written records—perhaps a name from a ship's registry at Ellis Island, an obituary in a newspaper, a name chiseled into a gravestone. But, at some point in the genealogical searching, the trunk of the family tree ends with nameless ancestors. Here genealogy turns into history, history into archaeology, archaeology into anthropology, and eventually our genealogical search digs down into paleontology and geology. The trail of ancestral evidence turns from specific ancestors to broad communities, to continents, to shards of bone that denote a prehistoric hominid species, to fossils in four-hundred-million-year-old rock that record some of the first footsteps on land, to molecular speculation about LUCA—the Last Universal Common Ancestor, an ancient single-celled organism that might bind together all animal life on Earth—and, finally, to billions-year-old rocks that record the Earth's emergence as a planet.

There are strong links between genealogy, evolution, and origins, yet in day-to-day dealings they feel tenuous, if they're felt at all. The words are often used interchangeably, though they can mean very different things. We tend to think of ancestors as our immediate predecessors, those whose names appear on a family tree or at least in the kind of family legend that gets hauled out at holiday dinners, as in "George Washington was a distant relative." In an evolutionary sense, we first look to the other primates as our closest genetic relatives, with whom we share distant relatives in ancient Africa. Then there are our origins, where evolution merges into a past so remote it feels more metaphysical than real. Yet all three—genealogy, evolution, and origins—fundamentally shape our personalities ("She's got her grandfather's temper"), our health (childbirth turned into a daredevil act by our mammalian propensity to give birth to live, large-brained babies), and ultimately by our very being as agglomerations of certain kinds of atoms and molecules that only together experience joy and anguish, as well as the mystery of sentient consciousness.

A common experience in genealogical research is that while digging into our past we discover that our origins are different than we'd thought. The true story of our past has been lost over time, neglected in boxes in attics and basements, or perhaps hidden. A black woman may discover a white great-grandparent; or the Muslim, a Jewish forefather; the adopted child may find that her mother was aboriginal. We don't always welcome these revelations. A new genealogical discovery connects us differently to the world. Perhaps a culture that was once fully "other" becomes part of "me." Maybe a distant land that was completely foreign and of no interest becomes a homeland. Someone in our neighborhood who we liked, or disliked, may become a distant relative. In gaining this knowledge, we are exactly the same yet somehow different. When we look in the mirror, we see ourselves a little differently. Suddenly there's new meaning in the shape of our nose or lips, the color of our hair or eyes. In knowing, we are given a choice: to accept this new view of ourselves or to deny, ignore, or reject it.

Genealogy is about the transfer of information, the genetic information that makes us who we are: tall, short, big-boned, sickly. This information is embedded in our DNA. We may think of DNA as the genetic information molecule, but DNA itself is a book made up of various words and sentences. Each of its atomic and molecular parts has its own characteristics in the ways it prefers to bond, in the energy of these bonds, in how it vibrates, in how it breaks. It's these chemical characteristics that make DNA what it is, that make us what we are. So, we ask: What shaped DNA? What is *its* heritage? It's the next step back in the family tree that takes us to the stars. DNA molecules, the fundamental building blocks of life, find their parentage, their essence, not here on Earth but in the broader cosmos. They are the language of the universe, the way the cosmos expresses itself. The information they inherited comes from the nature of the stars and the cosmic ecology that birthed them.

Many ancient cultures intuited an intimate, living connection between the heavens and the Earth. The cycles of the seasons were measured by the movements of stars and constellations at sacred sites, from Stonehenge in England to Machu Picchu in Peru. Women's cycles of waxing and waning fertility were in tune with the fattening and shrinking of the Moon. Many cultures worshipped the Sun and Moon as divine foreparents, father Sun and mother Moon. The Sun and Moon were out there, untouchable and holy life-giving forces on whom our lives depended for guidance regarding not only when to plant and harvest but also the success of that harvest. We ate because of the Sun. Today, this connection to the cosmos is largely seen as mythic, as stories that ancient peoples concocted to explain their existence.

Since Galileo first used a telescope in 1609, we've been on a voyage out, from the shores of the Earth into the cosmos. With our eyes, we've traveled farther into space and deeper into time, reaching the very beginnings of time: the big bang. Voyages often change the traveler. How have we been shaped by our collective journey, and what have we discovered? To my mind, one amazing realization is foremost: we thought we were exploring "outer space," but we are finding our deepest selves. The 1972 Apollo 17 image of the whole Earth from space—the photograph of the big, blue marble—deeply shaped our view of ourselves as planetary beings. This realization fueled both the environmental movement and a swell of planetary consciousness that's still building. It is one view of ourselves, as an isolated planet, alone in the blackness of space. The Stardust Revolution is providing a complementary image of humanity: not as alone and different but as intimately connected with—indeed, born of—the cosmos.

Through extreme genealogy, we've come to understand a new story of ourselves: the story of our origin in the stars. When we look at images of countless distant galaxies, glimmering planetary nebulae, or the shimmer of the Milky Way, we are looking at family photographs. We are separated by great distances of time and space and are not direct relations. We feel as different as we do from some multi-eyed, antennaed Cambrian creature preserved in the rock of the Burgess Shale. Yet, if we open ourselves to what we've discovered on our journey out, we come to see that the greatest discovery is not what's out there but rather how we're part of it. The guiding question of the Stardust Revolution isn't "Are we alone?" but rather "How are we connected?"

Ancient mariners looked to the stars to determine where they were. In the Stardust Revolution, we're realizing that the stars are guides to who and what we are. If we discover other life, it will not be alien; rather, it will be distantly related to us. We already find it difficult on Earth to understand our shared humanity, the deeply shared nature of black and yellow, white and red, man and woman—and we find it easy enough to detect aliens on Earth based on differences in culture, language, and nationality. This is the heart of the Stardust Revolution: it's as much about ourselves as it is about what's out there. We are stardust; the stars are our ancestors. When we look at the sparkle in a baby's eyes, we're seeing the reflected twinkle of long-ago stars shining once again.

NOTES FOR THE JOURNEY

Searching for our genealogical roots always eventually brings us to foreign lands, times, cultures, and languages. With the extreme genealogy of the Stardust Revolution, we quickly find ourselves in distant cosmic realms where the language that's used to describe it is that of astrophysics.

Taking an astrophysical genealogical journey is a trip, by Earth standards, into the land of extremes. You can leave behind your wristwatch, notions of a country mile, or thoughts of a cold day being one on which you see your breath in chilled morning air. On this trip, temperature, distance, and time are measured in ways we don't use in day-to-day activities. This is not to say that these conditions are somehow "other," but rather that, in understanding our cosmic heritage, we see that we embody astrophysical extremes.

To describe temperature, astrophysicists generally use neither the familiar Celsius nor Fahrenheit temperature scales but rather the Kelvin scale. The Kelvin temperature scale is calibrated in relation to absolute zero, the temperature at which atoms stop moving. On our genealogical journey, finding ourselves somewhere really cold—say, inside a dense molecular cloud from which solar systems are born—is to be at about 15 degrees Kelvin. That's approximately –433°F, or –258°C. For our purposes, all temperatures in this book are described in degrees Fahrenheit. I hope that for readers more familiar with the Celsius scale, –433°F will register simply as cosmically cold.

For the Stardust Revolution story, distances are often best described in light-years. A light-year is the distance a photon of light travels in one Earth year. Since light travels at the constant speed of about 300,000 kilometers per second, or approximately 186,282 miles per second, this distance is about 10 trillion kilometers or 6 trillion miles—huge numbers more simply described as a light-year. For comparison, the distance from the Sun to the Earth is about 8 light-minutes, or 93 million miles.

In this book, when smaller distances, sizes, or masses are involved, I've generally used the American Imperial measurement system of miles, inches, and pounds. However, when it comes to sizes and masses, I've used metric measurements when the original research did so, such as using nanograms or billionths of a gram to describe the mass of a grain of stardust. Again, I hope that for readers more familiar with other units of measurement the key message will be clear that something is either infinitesimal or enormous.

Cosmic genealogical time is a mix of the atomic-clock fast and the seemingly eternal. On the one hand, there are the atomic transformations in exploding stars that take place in billionths of a second; on the other, there are the interstellar wanderings of stardust that can take millions or billions of years before a single grain becomes the stuff of a star or planet. Nineteenth-century geologists tracing our evolutionary roots had to get used to talking about *deep time* measured in hundreds of millions of years. On our extreme genealogical search, pushing back our origins to before the existence of the Earth, a million begins to feel a little like spare temporal change, and you get to dropping the term *billion* like a Wall Street investment banker.

Now it's time for our journey.

PART 1

BORN OF STARS

CHAPTER 1

THE STARDUST REVOLUTION

We live in a changing universe, and few things are changing faster than our conception of it.

—Timothy Ferris, *The Whole Shebang*, 1997

MEETING LUCY ZIURYS

When I first met Lucy Ziurys in early December 2008, I was struck by the thought of what it would be like for a pre–World War II astronomer to meet her. He wouldn't believe she was of the same academic species. The time-traveling astronomer would think that he hadn't moved just through time but that he had also moved into a strange parallel universe. Ziurys's basement office in the Steward Observatory at the University of Arizona would at first seem familiar enough. Every horizontal surface—her desk, the meeting desk—is covered with piles of papers, a clutter reminiscent of that of legions of scientists for centuries.

But upon closer inspection and questioning, it would be clear to the visitor that this is a very strange future. For one, Ziurys is a woman, her shoulder-length, sandy-blonde hair is held back with a simple headband revealing a strong Nordic face with deep-set blue eyes. She has the direct bearing of a woman who has made her way in a field through dint of effort and will in what has squarely been a man's domain. A female astronomer with a prominent academic post, research funding, and a coterie of graduate students was unheard of until the latter part of the twentieth century. Yet our visitor might reflect that women had indeed played a major role in the laborious work of early twentieth-century astronomy and that Ziurys's occupation was thus not that strange.

Further inspection of her office would shake this initial small comfort. On her desk, under a traditional ball-and-stick model of a molecule common to chemistry classes worldwide, is a copy of a test from Ziurys's course, Astronomy 522. The test's first question asks students to draw the electron configurations for a number of molecules. Our visitor, probably wearing tweed, would wonder what an astronomer was doing teaching chemistry. This thought would only be confounded by a look at Ziurys's bookshelf, filled with dozens of multicolored binders, each binder labeled with molecular nomenclature that an astronomer with even an inkling of chemistry could see were alien combinations, and one simply titled "Extragalactic Molecules."

"Incredible!" our visitor would assert as he put the pieces together. Ziurys is both a chemist and an astronomer. She's an astrochemist, an astronomer who studies the previously unimaginable—not galaxies, stars, or planets but the molecules around and between stars light-years beyond our Solar System. Our visitor wouldn't be glimpsing just the future but also a new universe. Until the late 1960s, it was generally considered that space was simply too harsh a place for any but the simplest two-atom molecules to survive. Ziurys's bookshelves, however, hold the story of an utterly different cosmos.

When our visitor would finally ask Ziurys how she can possibly study cosmic chemistry—observing molecules so small they are too tiny to view with microscopes on Earth—the full impact of this bizarre future would hit home. Ziurys doesn't peer through a telescope to *look* for molecules in space; she *listens* for them. She's a radio astronomer. Hers is a perspective of the universe wholly unknown to astronomers before the advent of the first dedicated radio telescopes in the 1950s. Yet when she turns the Steward Observatory's Kitt Peak radio telescope toward any point in the sky, the signal-display monitor sings with the molecular signatures of vast seas of molecules stretching across the Milky Way.

By now our visitor, quite perplexed, might turn away and toward the windows for some intellectual relief—at least daylight would be the same. However, there on the ribbon of wall beneath a rectangular slit of window, he would perhaps see the strangest item of all: a framed certificate of appreciation from the US Space Studies Board. This would be fitting for an astronomer. But he would be perplexed by the committee of which Ziurys was a member: the Committee on the Origins and Evolution of Life. This astronomer isn't just a chemist; she has something to say about biology.

In 2004, when Ziurys was on the committee and the US space program was sagging from two space-shuttle disasters, President George W. Bush announced a "reinvigorating" national program of possible American missions to the Moon and eventually Mars. What the president didn't mention, or perhaps even know about, was a far grander space race already under way: an epic quest to see our cosmic selves anew. Eight years before, in a brief address to the US Congress, NASA's top administrator, Daniel Goldin, tried to galvanize lawmakers with a new vision of the American space program, a vision that didn't involve astronauts or manned spaceflight—at least not for the foreseeable future. It wasn't even ultimately about *out there*. It was about us. The program was Origins. "*Origins* is one of the boldest challenges NASA has taken on," Goldin told Congress, "and the results could literally change the way humans think about the universe and their place in it. . . . It will rewrite textbooks in physics, chemistry, biology, and quite possibly, history." Though he was the head of the US space agency, Goldin believed that the opening century of the coming millennium belonged to biology—that "the right stuff" in space research was about life. Goldin, and others in NASA's Office of Space Science, was inspired by the previous year's discovery of the first planet around a distant Sun-like star, an exoplanet. For millennia, such planets had been the stuff of myth, speculation, and, most recently, science-fiction movies such as *Star Wars*. Now the space science and astronomy community reacted like Alice having fallen through the rabbit hole—it had discovered a whole new cosmos, in which the notion of planets didn't end at Pluto. Goldin championed the Origins program as an interconnected weave that would extend from exploring the origins of galaxies to the origins of solar systems and finally to the origins of life itself. Tying the program together was a single guiding question: "Where did we come from?" Origins, it turned out, was too bold a program. The United States was still deeply mired in the debate over terrestrial evolution, one in which Congress had banned NASA support for the search for extraterrestrial intelligence (SETI) program it had pioneered, and one congressman had derided the program as "the search for little green men." Without the iconic appeal of manned exploration, Origins has yet to fully take off.

However, the new astrophysical discoveries that were driving Goldin's vision kept coming, fueling science at the intersection of astronomy and biology. At the same time that President Bush was announcing the Moon and Mars missions, the seventeen members of the Committee on the Origins and Evolution of Life were writing a landmark report joining astronomy and biology. Their report, *The Astrophysical Context of Life*, released in 2005, was routine in format and pedigree. But the question it asked and the conclusions it drew were historic. The authors reflected on one key question: What can astronomy tell us about biology and life? While the Bush administration was fighting the Culture Wars—tacitly, when not actively, opposing the teaching of terrestrial evolution in US schools—only several blocks away from the White House, at the offices of the National Academies, the members of the Committee on the Origins and Evolution of Life had come to the conclusion that the question of evolution on Earth was a twentieth-century issue. The twenty-first-century question was a cosmic one. Their report opened with the view that "there are compelling reasons to argue that a full and complete picture of the origin and evolution of life must take into account its astrophysical context."

The mix of biologists and astronomers who made up the committee were far from unanimous in their view of the scope or depth of this astrophysical impact. "I think everyone on that committee felt there was life elsewhere," says Ziurys the astrochemist in her basement office at the University of Arizona. "But there are people who felt that all life on Earth . . . evolved here in a soup on the Earth, with no connection out to space. And there are those of us that [ran] that committee who felt that there was a connection between what is occurring out in interstellar space and . . . how life evolved on Earth—or any planet." The sticking point for the committee wasn't evolution; it was the broader notion of origins. "You would be surprised how people think in silos, even in the scientific world," says Ziurys. "If they are biologists, working with Petri dishes in a laboratory, [they think the early] Earth was one big Petri dish."

For Ziurys, however, the ultimate key to understanding the origins and evolution of life on Earth *isn't on Earth*. To understand why all terrestrial life is carbon-based, why life uses only twenty of the possible dozens of potential amino acids, why iron is the metal atom around which the hemoglobin in our blood binds—to understand any of these life mysteries—we must look to the stars. For Ziurys, and a new era of scientists, our story doesn't begin on Earth; it begins with stardust.

THE THIRD GREAT REVOLUTION

Pick up a dictionary and look up *stardust*, or Google the word, and you'll see that it is culturally defined primarily as fantasy rather than as fact. *Stardust* is the title of novels, science-fiction movies, and Hoagy Carmichael's 1927 hit song—one of the most popular American tunes of all time. Most dictionaries' definitions of *stardust* are similar to this one from *Merriam-Webster*: "a feeling or impression of romance, magic, or ethereality." Stardust is the equivalent of fairy dust—the stuff of fantasy, intangible and elusive.

But we're in the midst of a cultural and scientific shift, and at its heart is the new science of stardust. It's captured evocatively in

NASA's Stardust mission, which in 2006 became the first mission to bring back samples from a comet: 242 million miles from Earth, the *Stardust* robotic probe intercepted comet 81P/Wild 2, swept through the plume of dust and water that make up the comet's translucent tail, and made it back to Earth with an invaluable cache of microscopic grains. Some of these tiny grains are literally stardust. Wild 2 was formed from a birth cloud of dust and gas that gravitationally collapsed to form the Sun, the Earth, and other planets, as well as the countless asteroids and comets that compose our Solar System—and, ultimately, you and me. The tiny grains of sand collected from comet Wild 2 have been largely unchanged for 4.5 billion years, from the time when the Earth was forming. Stardust is now the stuff not only of fantasy but of fact and science.

Stardust science isn't a term you'll find in scientific journals. I've coined it for two reasons. First, it captures a profound shift in our understanding of the world and the cosmos. Second, it envelops the scope, majesty, and essence of a diverse range of research, all linked by literal stardust. Stardust science has developed gradually and spasmodically at the peripheries of the established departmental sciences of astronomy, physics, chemistry, geology, and now biology. Many scientists engaged in this work don't attend the same scientific conferences. In the increasingly fragmented, niche-specific realms in which they work, they often can't understand the details, or the importance, of each other's scientific papers. They lack a common language. For example, when biologists talk about *extinction*, they mean the elimination of a species. When astronomers use the term *extinction*, they're referring to the degree to which matter between the stars blocks their view of light from a distant star or galaxy. But indicative of the emergence of a new science, other scientists are bridging these linguistic gaps and creating interdisciplinary understanding. In the process, they're reversing a two-hundred-year trend toward increasing scientific fragmentation.

At its core, stardust science is perhaps science's greatest exercise ever in integration, extending the notion of ecology into the cosmos.

German zoologist Ernst Haeckel coined the term *ecology* in 1873 to refer to the new science dealing with the relationship of living things to their environments. He developed the word from the Greek *oikos*, for "house," and *logia*, "study of." Thus, stardust science seeks to find our home in the greatest environment of all, the universe. As the *Astrophysical Context of Life* report contends, for example, the study of the molecules of life on Earth "should be connected to topics of star formation and cosmochemistry and the origin of life." At the heart of this integration is the new field of astrobiology, the great unifying science. It draws not just on astronomy and physics but also on chemistry, biology, and planetary geology, extending all these disciplines to the stars in the search for life's cosmic origins and connections. Astrobiologists aren't probing the universe's physical structure but rather its biological nature.

Historically, astronomy and biology are the strangest bedfellows. Those who studied cells didn't study stars, and vice versa. One group looked down through microscopes into the essence of our beings; the other looked up and away through telescopes into the depths of the cosmos. But through this searching of inner and outer worlds, some biologists and astronomers have sensed a common goal and a single connected story. The old, seemingly impenetrable wall between our evolutionary nature on Earth and the cosmic story we see around us is crumbling. Evolutionary theory is entering the space age. For stardust scientists, the focus isn't on elucidating an expanding universe but rather on an evolving one. This is not the view of a scientific fringe. It is captured in the words of another landmark document, the 2008 *NASA Astrobiology Roadmap*: "We must move beyond the circumstances of our own particular origins in order to develop a broader discipline, 'Universal Biology.' . . . We need to exploit universal laws of physics and chemistry to understand polymer formation, self-organization processes, energy utilization, information transfer, and Darwinian evolution that might lead to the emergence of life in planetary environments other than Earth."

Through the research of scientists like Lucy Ziurys, we are in the midst of the third in a series of scientific revolutions that have shaped our understanding of our origins and place in the cosmos. The first revolution was the Copernican Revolution, which in the sixteenth and seventeenth centuries removed the Earth from its divine locus as the center of creation and joined our planet with the other planets orbiting the Sun. Three centuries later, the Darwinian Revolution removed humanity's distinct, divine biological status to place this species in the ebb and flow of all life on Earth. We are now in the midst of a third seismic shift in our understanding of our place in the living cosmos—the Stardust Revolution. It is merging the Copernican and Darwinian Revolutions, placing life on Earth in a cosmic context.

THE ORIGINS OF THE STARDUST REVOLUTION

Stardust science emerged as the unsuspected offspring of twentieth-century astrophysics, the marriage of astronomy and physics. Astronomy in the twentieth century was dominated by the question of the physical origins of the universe. Nineteenth-century astronomers had looked up at a heavens that had no known age, size, shape, or origin. The night sky was a dark well of the unknown. Twentieth-century astrophysicists, with new telescopes and the tools of physics, have performed the greatest pull-away dolly shot in history. They've moved the camera back to reveal Earth not just as the third planet from our Sun but also as a planet nestled among billions of stars in the spiral arm of a galaxy, the Milky Way. Then, in epic fashion, the astrophysics camera pulled back even farther to give us a divine perspective of a cosmos that evokes gasps as we see billions of galaxies in an infinite universe. Today you can download to your computer monitor the latest deep-space image from the Hubble camera, confident that the universe is 13.7 billion years old (give or take 0.13 billion years) and that it will continue to expand forever.

This comprehension required great intersections of scientific theory, observation, and experimentation. These included Einstein's general theory of relativity, laying out a basis for the nature of space-time and cosmic-scale gravity; Edwin Hubble's meticulous measurements of the speed and direction of distant galaxies, revealing their increasing separation and an expanding universe; and Arno Penzias and Robert Wilson's serendipitous discovery of the cosmic microwave background radiation—a universe's birthing sounds, evidence of the big bang. The universe revealed by astrophysics is full of exotic objects: exploding stars that for months are brighter than an entire galaxy; neutron stars, stellar remnants so dense that a teaspoonful of this über-compact matter weighs more than all the buildings in Manhattan; and, at the hearts of galaxies, massive black holes—objects whose powerful gravity traps even light.

Taken together, it's an astrophysics story that culminates in the present-day Standard Model of Cosmology, the scientific theory that explains the origins and structure of the universe. This theory ties together particle physics and cosmology, and predicts the existence of, as yet invisible, dark matter and energy, thought to account for about 95 percent of cosmic mass. It provides the scientific shoulders from which physicists such as Stephen Hawking reach for a unified theory of physics, one that joins the theory of the biggest objects, general relativity, with those of the smallest, quantum mechanics. Yet, as Hawking concludes in *A Brief History of Time*, even this achievement wouldn't close the circle. "A complete, consistent, unified theory is only the first step," he writes. "Our goal is a complete *understanding* of the events around us, and of our own existence." How do *we* fit into this story of the universe?

The puzzle of our cosmic origins is the great untold science story of the past five centuries. It's been at the core of work by scientists whom we know for other work: Newton's monumental insights on gravity and light; Pasteur's legendary experiments on the microbial basis of disease; Bunsen's iconic chemistry burner; and Einstein's defining theories of space and time. It was there, sometimes as a secret, at other times as a surprise or beyond the realm of experimentation, at the core of most of the great scientific debates—evolution, the spontaneous generation of life, the origin of the universe, the nature of stars, the origin of atoms. Each debate provided an odd-shaped piece in a great jigsaw puzzle whose assembled pieces now reveal the outline of an incredible image barely hinted at by any one piece. This is a cosmic puzzle for which we haven't had a box-cover image to guide our assembling. We've had to grope, feeling the shape of each piece, finding edges that fit—or that seemed to fit—only later to be dramatically taken apart and rearranged, and then often with our forgetting that we'd ever arranged them otherwise. All along, we thought that this puzzle was about the world out there, about Nature or the cosmos, but not about us. Yet in that amazing tradition of outward journeys in which we find ourselves, we've created an image in which we are now faced with our own reflection.

NEW WAYS OF THINKING

Stardust science seeks to place the Earth and our existence in a historical, cosmic context. For stardust scientists, the question of biological evolution doesn't begin with the creation of the Earth but pushes back in time and space. In cosmic genealogy, the Earth is a recent arrival, only a third the age of the universe. How do we understand the origins of life in a cosmos brimming for billions of years before the Earth existed? When astrophysicists discovered the first molecules in outer space in the late 1960s, the scientists who studied them gave themselves the mongrel title "molecular astrophysicists." They thought in physics terms about these molecules light-years away—in terms of their quantum energy states and the wavelengths of light they emitted when energized. But during the 1970s, a new breed of astronomers looked at the molecules between the stars not as physicists but as chemists—astrochemists. It was a profound difference of perspective. Where physicists see stars as objects that produce light and heat, astrochemists see stars as the source of elements and molecules, and, therefore, ultimately, life.

The prophet of this new way of looking at the heavens was the American astrophysicist turned astrochemist Carl Sagan. Sagan was above all a critical thinker. While he inspired millions as one of the great astronomy popularizers of the twentieth century, he also sought to marry this new view with new ways of thinking. His book *The Demon-Haunted World*, for example, explored the boundaries between superstition and science. His critical thinking as an astrochemist and astronomer led him to a singular conclusion: it's not just the Earth but the entire universe that's alive. The opening sequence in Sagan's PBS television series, *Cosmos*, evokes this verdant view. Sagan, standing atop a cliff on the wind-blown coast of a Hawaiian island, introduces viewers to a new story: rather than being a barren, inanimate desert of high-energy particles and black holes, the cosmos is a fertile sea awaiting our discovery. As Sagan wrote in the companion book to the TV series: "The surface of the Earth is the shore of the cosmic ocean. . . . Recently we have waded a little out to sea, enough to dampen our toes or, at most, wet our ankles. The water seems inviting. The ocean calls. Some part of our being knows this is where we came from."

For all its poetry, and in some cases because of it, Sagan's message lacked convincing substance for many in the scientific community. Among his many colleagues, Sagan's vision was fundamentally undercut by one fact: in 1980, we knew of only nine planets (including the since-demoted dwarf-planet Pluto) not just in our Solar System but in the entire cosmos. Extraterrestrial life needed more than the molecules of life; it needed somewhere to live. As far as we could tell, however, the other planets of our Solar System lacked a heartbeat. They were lifeless orbs of rock and gas. But in December 1996, fifteen months before Sagan's death, the extraordinary discovery of planet 51 Pegasi b (commonly referred to as "51 Peg b") broke a scientific spell—the discovery of the first exoplanet, a planet around a distant Sun-like star. For half a century, almost all astronomers believed that if there were planets around other stars (and many doubted this), they would be impossible to detect, let alone study. It took thousands of hours of telescope searching, technical innovation, personal and professional heartbreak and disappointment, and endless errors to discover this first planet orbiting a star other than our Sun.

The discovery of exoplanets did more than launch the greatest search of all time. It was the breaking down of the Berlin Wall of the Stardust Revolution. Change had come. And with it, a new way of thinking took hold—that the cosmos' biological nature is knowable. For decades, astronomers, including Carl Sagan, who talked about the study of life in the universe, were derided by scientific colleagues for pursuing a topic without a subject matter. Beyond Earth, in our Solar System there was no other life, and, more importantly, there was no way of studying the question of life in the universe or its cosmic origins. End of discussion. But the discovery of exoplanets was the tipping point in a decades-long buildup of evidence—from the discovery that the elements are forged in stars, to the discovery of interstellar water and organic molecules and of amino acids in meteorites—that is the foundation of the Stardust Revolution. The biological universe could now also be described by Einstein's statement of astonishment: "The most incomprehensible thing about the universe is that it is comprehensible."

NEW WAYS OF SEEING

In the Stardust Revolution, this growing comprehension is driven by new ways of seeing the cosmos. The history of science is the history of resolution, and this is no more so than with stardust science. With each advance in telescopes, we don't just see better and more, we see differently. The more detail we can see, whether in a cell or in the cosmos, the more we know. Until Galileo looked through his telescope, there was no way of seeing Jupiter's moons, no way of seeing that the milkiness of the Milky Way is in fact the light of countless individual stars. The lenses in Galileo's telescope, crafted with what was then the world's finest glass, made by Venetian glassmakers, provided only about thirty times magnification, but the impact on human consciousness was tremendous and far reaching.

The Stardust Revolution involves not just greater resolution but different ways of seeing. Prior to World War II, humanity had explored the heavens only in light visible to the human eye, but today optical telescopes are only part of an armada of observatories, many of which don't see in visible light. Stardust scientists are exploring the cosmos in almost every wavelength of the electromagnetic spectrum, from gamma rays to x-rays, ultraviolet to infrared, and microwaves to radio waves. Each wavelength reveals different aspects of cosmic nature that are often impossible to observe in other wavelengths. It took the serendipitous 1940s development of radio astronomy—of using not visible light but radio waves and microwaves—to explore the heavens and reveal the previously invisible molecular universe. Space-based infrared telescopes, such as NASA's Spitzer Space Telescope, reveal the dusty origins of distant stars and planets. Future ground- and space-based telescopes will probe the atmospheres of Earth-like planets around distant suns.

BEYOND THE IMPOSSIBLE

The Stardust Revolution is also the story of what it's like to live in revolutionary scientific times. It is a revolution in the sense of what science historian Thomas Kuhn called a "paradigm shift," in which there is a gradual building of evidence against the predominant worldview until one day, while that view still seems strong, the scientific old guard lays down its arms and the wall comes down. Revolutions in science involve all the elements of political revolutions, at least bloodless ones. Sir Fred Hoyle, the British astrophysicist who in many ways was the great twentieth-century prophet of stardust science, noted in his autobiography that "I grew up with the erroneous notion that the scientific establishment welcomes progress, which is the opposite of what is generally true. Progress is equivalent to revolution." There are opposing camps: one entrenched in the bastions of power, the other claiming that the emperor has no clothes. It might seem like a tempest in a teapot, and at times it is, but there's something much larger at play—our understanding of ourselves, the universe, and our place in it. In science, the revolutionaries fight to get funding, to get published, to get telescope time, and more. Their scientific papers are often rejected and even ridiculed by anonymous reviewers. Stardust revolutionaries have faced an establishment that has dissuaded and ridiculed their research; that has denied, dismissed, or simply ignored findings.

Scientists have long had their own way of describing the painful and frustrating experience of a paradigm shift. Fred Hoyle's colleague, the Cambridge University physicist Ray Lyttleton, said that there are three stages in the acceptance of upstart scientific ideas: first, the idea is nonsense; then, somebody else thought of it before you did; and finally, we believed it all the time. The Stardust Revolution has been a centuries-long dance between what seemed reasonable at the time and what observation actually revealed, between elaborate theories of how the cosmos worked and singular theory-shattering discoveries.

In retrospect, we see a single narrative. In many cases, there's a stardust science pedigree of scientists trained and inspired by stardust mentors who in turn go on to train and enthuse a next generation. It's not, however, lived as a single narrative but rather as a maelstrom of uncertainty in which there are countless possible endings to the story. For all its messiness, though, science gropes and stumbles toward observable truth. Many of the establishment's leaders became converts over time. Though resistant at first, they are drawn to a vision because they refuse to deny the truth of their own scientific intuition and observations. The stardust story is possible because of the strength of their personal convictions, their wide-ranging curiosity, and their vivid imaginations. Because a universe seeking to know itself urged them to speak their truth.

The modern space age began in 1957 with the successful launch and orbiting of the Soviet Union's *Sputnik 1*, the world's first satellite.

Sputnik sparked the frantic creation of NASA a year later and also the US National Research Council's Space Studies Board, which Lucy Ziurys would later join. But 1957 was also the year that Fred Hoyle and three colleagues wrote a seminal paper, one of the landmarks in the history of science, describing how stars fuse hydrogen and helium to create all the other elements in a process called stellar nucleosynthesis. Thus, they showed that with every atom of carbon, oxygen, iron, and calcium in our bodies, we are truly the stuff of stars. While the world's attention was occupied by its fear of intercontinental missiles and the race to the Moon, the Stardust Revolution had begun.

Today, after millennia of stargazing speculation and nearly a century of space-age science fiction, the search for extraterrestrial life is real. When humans next voyage to the Moon, we might do so with the knowledge that light-years away is another living world; that on a distant exoplanet—under the light of a full moon, or two—other beings experience lives of mystery; that not only are we not alone, but these beings are our relatives, made from the same materials and as a result of similar biological processes; and that just with every atom of our bodies, we are joined with a living cosmos.

Stardust science raises profound questions about previously staunch intellectual boundaries between living and nonliving, between us and everything else. How did life begin before the Earth took shape? What's our living relationship with the rest of the universe? Who and what are we, in a cosmic sense? What is life? Are there abundant Earth-like exoplanets, as is argued by many astrobiologists, or are we an exquisitely "rare Earth," a singular blossom in an otherwise utterly barren cosmos? At its core is one question: Is life a fundamental emergent property of the universe?

To tell the story of the Stardust Revolution, it's necessary to start at the beginning, at a time when astronomers and physicists were struggling to explain the basic nature of stars. As so often happens when we travel, in seeking to understand something else, what is most profound is what we discover about ourselves.

CHAPTER 2

A STAR'S FINGERPRINT

The tiny, twinkling stars of the night sky, shining with feeble and fluctuating beams, appear so minute and unsubstantial that an inexperienced beholder might expect one after another to vanish from its place. . . . Instead of infinitesimal points of light . . . [the astronomer] visualizes them as giant globes of incandescent matter pouring forth energy at so terrific a rate as to stagger the imagination. The contrast between the apparent and the real is the most stupendous in all human experience.

—Paul W. Merrill,
"Stars as They Look and as They Are," 1926

LOOKING AT THE SUN

Every workday, Steve Padilla gets up and does what mothers the world over tell their children not to do: he looks at the Sun. Atop Mount Wilson, 5,700 feet above the Los Angeles basin, from where you can only imagine its byzantine freeways snarled with traffic, Padilla leaves the modest bungalow that he's called home for the past twenty-five years and makes his daily pedestrian commute under the towering Douglas fir and the majestic canyon oak to his solitary job at the 150-Foot Solar Tower. This observatory doesn't look like the famous dome-shaped observatories Padilla passes on his way to work. The Solar Tower resembles a Texas oil rig more than it does a telescope. Its four steel-girdered legs surround a central tube, with the addition of a small dome on top. But this structure isn't mining down for crude; rather, it's looking up for sunlight. Padilla climbs into the rickety, open-bucket elevator on the tower's side (unless there's a wind over sixty miles an hour, in which case he should use the ladder, though he never has) and makes a clanking, slightly swaying ascent up to the dome. From here he gets the best view of the approximately seven million souls in the greater Los Angeles area. It's a vista that captures the contrasts of Southern California: to the west, he looks down to the San Gabriel Valley and the seemingly endless spread of the City of Angels a vertical mile below; to the east, the Sun rises over a tiered wilderness of snowcapped mountains.

Padilla opens the observatory dome and positions the primary mirror so that it aims the dawn light directly at a secondary mirror, which in turn reflects the soft morning light down into the observatory's heart. For the rest of the day, the mirror will automatically track the Sun's journey across the Southern Californian sky.

After returning to the observatory's viewing room, Padilla, monk-like, with thinning curly hair and a ripped blue windbreaker, gets a pencil and positions a fresh ten-by-twenty-inch piece of drawing paper at the light focus so that the orb of the Sun is centered in the middle of the sheet. Then, as Mount Wilson's solar observer has done for most of his life—certainly longer than any other person on Earth—Padilla draws by hand the location of sunspots. These slightly cooler, magnetized areas of the Sun's surface appear as dark blemishes on the solar face. On this day, April 14, 2011, he marks that the *seeing*, the term astronomers use to describe the clarity of their view, is 2.0 out of 4—a little hazy, the spring air still turbulent from a high front moving in, clearing out the previous day's clouds.

Padilla started as the solar observer in August 1976, landing the coveted post after first working as a relief night assistant on Mount Wilson's sixty- and hundred-inch telescopes. For thirty-five years he's been observing our star, recording its eleven-year cycles of dipping and peaking magnetic activity, clocking its rotational speed, and measuring its overall intensity. While for most of us the Sun is a constant, for Padilla "there's always something new." What's new is usually the distribution of sunspots. During years of low sunspot activity, the Sun can go spot-free for months. Five years later, at the sunspot peak, the Sun's face can resemble that of an acned teenager, marked across its diameter with dozens of pairs of spots, each a north or south pole to the other.

The sunspots are like islands on the Sun's surface that appear to move across its face as the fiery orb spins on a twenty-seven-day rotation—a solar day. For the past several years, Padilla hasn't taken long with the drawings; sunspot activity has been on the wane. He's faithfully recorded the downward slope in the Sun's eleven-year sunspot cycle of activity, a process he's watched personally for a full three cycles. Today there is a smattering of sunspots along the Sun's 20° north line of latitude, and in the next eleven-year cycle the spots will mysteriously flip to the Southern Hemisphere. The largest sunspot visible today, a dimple on the solar surface, is about one and a half times the Earth's circumference. Padilla shows me the image of the largest-ever recorded sunspots from the same month in 1947, when the spots were massive, like small seas of darkness dozens of times larger than today's spots.

Padilla finishes his drawing and puts it into a binder for future reference. His drawing is the latest in an unbroken run of about thirty thousand daily Sun sketches made at the 150-Foot Solar Tower since January 4, 1917, and is now continued more as a labor of love than for science. Padilla is part of a dying breed, a ground-based solar observer. Today, most Sun observing is done with robotic telescopes or space-based satellites.

Padilla's early-morning isolation belies a broader truth about the observatory's busy, central role in astronomy history: Mount Wilson is the site of the most important astronomical discoveries of the first half of the twentieth century. On the walls of the Solar Tower are pictures of some of the famous visitors who've made the pilgrimage up the mountain, including Albert Einstein in 1931, on one of several visits, and Stephen Hawking, his thumbprint marking the visitors' book from June 2, 1990. It was here, atop Mount Wilson, that Albert Michelson accurately measured the speed of light; that Harlow Shapley determined our Solar System's position in the suburbs, rather than at the center, of the Milky Way. But what draws legions of visitors now is Mount Wilson's legendary status as the site of Edwin Hubble's astronomical insights. In 1925, using Mount Wilson's one-hundred-inch telescope—the world's most powerful at the time—Hubble discovered that distant "nebulae," the previously mysterious smudges of light on the inky night sky, were in fact vast agglomerations of stars—galaxies. The upshot was that the Milky Way wasn't the entire universe but rather an island of stars amid millions of others. In 1929, Hubble announced that his observations showed that these galaxies are moving apart, and that the farthest ones are moving apart faster than the nearby ones. He'd shown that the universe wasn't just vaster than anyone imagined; but that it was growing.

What's often lost in this best-known version of Mount Wilson's story is that before the nighttime telescopes arrived, the observatory was established to do astronomy under the full glare of the Sun, with the goal of understanding our evolutionary connection to it.

THE GREAT SEER

Silently watching over Padilla, in the form of a ceramic bust set atop a red mechanics tool cabinet in the solar observatory, is the original Sun watcher himself, George Ellery Hale. No one has had as great an impact on modern astronomy as Hale. Without Hale, there might well not have been Hubble or the dozens of lesser-known astronomers who laid much of the foundation of twentieth-century astronomy and cosmology. Born in 1868, as the United States was still reeling from the Civil War, Hale was the son of a Chicago elevator entrepreneur who made a fortune installing "vertical railways" in the Windy City's burgeoning sky-reaching buildings following the Great Fire of

1871. But Hale Jr. lifted humanity to greater heights. He didn't just establish Mount Wilson, he also envisioned and built an astronomy empire. Founding just the Mount Wilson Observatory would have been a life's work for any ambitious scientific visionary, but Hale was also seminal in founding the California Institute of Technology, the United States National Research Council, and the International Astronomical Union. Hale was the scientific equivalent of the legendary American industrialists, including John D. Rockefeller Sr. and Andrew Carnegie, whom he pursued and convinced to fund an unprecedented journey into the heavens.

Like the titans of industry, Hale was dedicated to the great American dictum that bigger, and technologically more advanced, is better. In astronomy, this is often the case. Like a pyramid builder, Hale could imagine projects that would take decades to complete and that would push human creativity into the realm of the seemingly impossible, but when they were completed, they would stand without equal. The observatories he built included the four biggest telescopes of the twentieth century. Each of these eyes on the universe outsized its predecessor; each finished telescope revealed ever farther and fainter glimpses of the cosmos. First came the forty-inch mirror at the University of Chicago's Yerkes Observatory in Wisconsin, then the sixty-inch telescope at Mount Wilson, followed by the observatory's hundred-inch telescope, through which Hubble made his grand discoveries. Hale's final triumph bears his name, the two-hundred-inch Hale Telescope at Mount Palomar, conceived in 1928, which saw first light in 1948. The five-hundred-ton Hale Telescope, with a curved mirror seventeen feet across, smooth to within the width of a bacterium (two-millionths of an inch), was Earth's greatest eye into the heavens until 1993, more than a half century after Hale's death in 1938. Hale's level of creative output was fueled by what today might be deemed bipolar disorder: bursts of activity interspersed with periods of profound depression and debilitating headaches that led to an early retirement.

What consumed Hale's deep passion and fed his workaholism weren't just the stars but also the deep belief that in them we'd learn something essential about ourselves. Hale founded the Carnegie Observatories in Pasadena, just ten miles from the mountain's base, as the location of offices for the Mount Wilson astronomers. In his office there, Hale's personal library, which included original works by both Galileo and Copernicus, also included a thumb-worn fifth-edition copy of Charles Darwin's *On the Origin of Species*. For Hale, evolutionary theory was as much a commentary on the origins of stars as of species. "It is not too much to say that the attitude of scientific investigators toward research has undergone a radical change since the publication of *Origin of Species*," he wrote in his 1908 book *The Study of Stellar Evolution*. "This is true not only of biological research, but to some degree in the physical sciences. Investigators who were formerly content to study isolated phenomena, with little regard to their larger relationships, have been led to take a wider view."

Hale believed that the key to this wider view of the cosmos began not with peering at distant stars but in studying just a single nearby one: the Sun. "We are now in a position to regard the study of evolution as that of a single great problem, beginning with the origin of the stars in the nebulae and culminating in those difficult and complex sciences that endeavor to account, not merely for the phenomena of life, but for the laws which control a society composed of human beings. Any such consideration of all natural phenomena as elements in a single problem must begin with a study of the Sun, the only star lying near enough the Earth to permit detailed investigation."

The idea of doing astronomy in daylight is counterintuitive until we remember that the Sun is in fact a star and not so different from the approximately two hundred billion other stars in the Milky Way or the uncountable trillions of stars in the universe. It was this insight that made Hale not an astronomer but a solar observer. From nineteenth-century discoveries, Hale knew that the study of sunlight could provide powerful clues to unlock the secrets of the night sky, a new field of study he called astrophysics. His observatories would be laboratories, places where astronomy met terrestrial physics; where, through an understanding of what we see, measure, and describe on

Earth, we can fathom the stars and in turn learn about ourselves. With astrophysics, astronomers could move beyond understanding *where* the stars are to knowing *what* they are. The key to this understanding lay in the Sun.

In 1903, having convinced Andrew Carnegie and his Carnegie Institution to pay the tab, Hale founded the Mount Wilson Solar Observatory. He chose this Southern Californian summit for its good "seeing"—the air masses over the peak were unusually quiescent—Hale arranged for Mount Wilson's first telescope, its sixty-foot solar telescope to be hauled up to the peak, mules carrying the telescope, every steel girder, and eventually the finely polished mirror, up from the dry, sage-scented canyons below, along precipitous and narrow cut-back trails today walked by weekend hikers packing electrolyte-balanced energy drinks. The solar observatory was completed in 1907, and a year later, Hale used it to become the first person to see that sunspots contain magnetic fields. Each spot is actually a married pair—a negative and a positive pole. It was a monumental insight, the first observation of magnetism beyond the Earth, adding one more astrophysical link between the heavens and the Earth. In order to tease more information out of sunspots, however, Hale needed a bigger telescope. Thus was born the 150-Foot Solar Tower, completed in 1912.

Standing in Hale's solar observatory while the repetitive whir of the magnetograph scans across the disk of the Sun, automatically measuring its motion and magnetic fields, I ask Padilla what he likes most about his job. "Every day I get to look at a star," he says with an impish grin, echoing Hale's vision. "It's our closest star. By looking at the Sun we're seeing how a star works." It's a line he's delivered to thousands of visitors, but it's no less powerful for that: strange as it is on one level, it's easy to forget. On clear nights, if we're outside the shroud of city lights, we see stars as twinkling points of light—deeply distant, cold, and mysterious. By midday, given that the stars now seem as though they have never been in the bright-blue sky, it might be difficult to believe that the intense, fiery, blinding Sun is just like those points of midnight light.

I've come here to see the solar spectrum. We see this spectrum in a dilute form when droplets of water in the atmosphere split sunlight to create a rainbow—dividing sunlight into its component parts from red through yellow to violet. But, using a spectroscope, the equivalent of a high-tech prism, the solar observatory dissects sunlight to create an enormously high-resolution spectrum, revealing otherwise invisible details of the stellar rainbow. Padilla adjusts "the periscope," a viewing tube for looking at the spectrum. I glance toward the wall at a picture of Einstein bent over, his eye on the periscope, and Hale bent over Einstein's shoulder, as if asking, *Can you see them?*

"There, take a look," says Padilla. I look through the periscope and gently move the view across the solar spectrum. There they are: the Fraunhofer lines, an intermittent series of seemingly randomly spaced thin, dark lines across the pale-green section of the spectrum Padilla has focused on. Not fuzzy or flickering, these dark lines are ruler-straight, evoking the dark lines drawn by a government censor to redact a document. Directly in front, I see a pair of lines, one a little thicker than the other; to the left is a thin, lighter line. Looking up, I see that these lines correspond exactly with the photograph of the Fraunhofer lines that Padilla uses as a benchmark to calibrate his observations. These particular lines are from atoms of ionized iron dancing in the Sun's atmosphere, absorbing photons of sunlight produced in the Sun's core, and thus acting as tiny shields revealing the Sun's composition. These Fraunhofer lines are the reason that Hale built this observatory. After the telescope itself, these lines are the cornerstone to our understanding of the cosmos. They were the great astronomical riddle of the first half of the nineteenth century. The serendipitous solution to that riddle set in motion a revolution: the Stardust Revolution.

OUT OF MYSTERY

Why did Hale so fervently think that studying the Sun would reveal the hidden nature of all stars? To answer this question, we have to go back to a book published in installments between 1830 and 1842: Auguste Comte's six-volume treatise *The Course in Positive Philosophy*.

Comte, one of the great nineteenth-century philosophers, was an eminently reasonable man. In fact, he was one of the most rational men of the century. Born in 1798, a child of the aftermath of the French Revolution, Comte expounded that a deliberate mixture of love and reason could form the basis for a new, harmonious social order, one that spurned the guillotine and rejected blind obedience to Church and monarch. Comte wrote prolifically, and his ideas were heard around the world. Brazil's fathers of independence were guided by his famous creed "Love as a principle, the order as a foundation, and progress as a goal," weaving a shortened version into Brazil's flag as "*Ordem e Progresso.*" Comte's fame waned in the past century, though the intellectual fields he founded would impress anyone. In his application of science and reason to human affairs, Comte inspired sociology, and his thinking about the sciences made him the first modern philosopher of science.

Comte wasn't merely interested in what we know but also in thinking about how we know things and, ultimately, what is knowable. In *The Course in Positive Philosophy*, he covered all the sciences of his day in incredible detail, from astronomy to biology. When this French philosopher looked at the stars in the night sky over Paris, he was certain of one thing: we would never know their deep nature. "Men will never encompass in their conceptions the whole of the stars," he reflected. "We can never know anything of their chemical or mineralogical structure."

Comte wasn't simply a naysayer. His statement came at a historic tipping point—the birth of modern science. He wanted to set a new standard for describing the world we see and our expectations of what we might come to understand. His point wasn't that the stars are innately mysterious or divine and therefore beyond human understanding. Comte's argument was based on deliberate reasoning. He pointed out that, then as now, we have three modes of scientific exploration: direct observation, experimentation, and comparison of similar systems in order to see commonalities and differences.

As for the stars and planets, he reasoned, "Experiment is, of course, impossible." You couldn't sample a star and bring a fiery chunk into a laboratory to study it under a microscope or subject it to the growing wonders of nineteenth-century chemistry and physics. As for comparison, he reasoned, it "could take place only if we were familiar with an abundance of solar systems, which is equally out of the question." Thus, he concluded, all that would ever be possible was observation. All you get from a star is a twinkling speck of light. Pretty, but not the stuff of science. For all the grandeur of the night sky, Comte believed that astronomy—limited as it was by the observation of faint light from distant bodies—was forever restricted to the domain of the mathematician, "measuring angles and computing times of the heavenly bodies." In the vernacular, astronomy was a dead-end discipline, but Comte reasoned that "if the knowledge of the starry universe is forbidden to us, it is clear that it is of no real consequence to us, except as a gratification of our curiosity."

He was far from alone in his assessment. Astronomers of the day could observe the motions of stars and planets but could not figure out what they were made of. Yet even as Comte was writing, the first cracks were appearing in the wall of the "impossible" nature of the stars. Within several years of Comte's death in 1857, scientists would proclaim that, far from being unknowable objects, it was indeed as if you could hold a star in your hands and tease secrets from it, a revelation that came not from looking at a shining star but at the light of a little flame.

BUNSEN'S BURNINGS

If there's just one thing generations of students remember from high school chemistry, it's Robert Bunsen's last name, immortalized in the eponymous small, metallic burner that's emblematic of chemistry class. It may come as a surprise to many familiar with Bunsen as a chemist that when he died at the age of eighty-eight on August 16, 1899, his obituary appeared not only in chemistry journals but in the *Astrophysical Journal*. Listing Professor Bunsen's many scientific accomplishments as one of the great chemists of the nineteenth century, the obituary writer reflected that "the work for which Bunsen will, possibly, be longest remembered is that which he did with Kirchhoff in establishing the science of spectrum analysis." Ironically, Bunsen is remembered for the burner he neither solely invented nor claimed as his own. Yet his deepest contribution to science is a largely unknown story. It was what he did with his burner—and the purpose for which he finessed it—that makes his the first great story of the Stardust Revolution.

Robert Bunsen had an early penchant for studying things that burned or blew up in the most violent ways. By all accounts, he was a gracious friend and colleague, but when it came to experiments, he loved to live on the edge. Bunsen's fame came not from great theoretical insight but from the fact that he was a persistent, deeply creative experimenter and laboratory innovator. He sought new ways to tackle vexing problems, even at great personal risk. In 1847, Bunsen traveled to Iceland, where he became fascinated with the island's angry geyser eruptions of searing hot water and steam. To understand their cause, he positioned himself on the lips of geysers and lowered thermometers into the geyser tube immediately before an eruption. In the seemingly safer confines of the laboratory, he discovered the first member of a series of arsenic-based compounds called cacodyls, named from the Greek term meaning "evil-smelling matter." To appreciate just what an accomplishment this was by nineteenth-century chemistry standards, when the idea of a fume hood was science fiction, you need to consider the substance's particular qualities. Cacodyl's smell resembles that of garlic, according to those who have smelled it—and lived to tell the tale, for it's also extremely toxic. Bunsen found that a vaporized speck of the stuff, today known as tetramethyldiarsine, was enough to kill a frog. Additionally, it spontaneously combusts in air. On one fateful occasion, a cacodyl explosion permanently blinded Bunsen's right eye and left him on the verge of death for several days from arsenic poisoning. Yet he seemed to revel in the substance's extreme noxiousness, noting that not only did it smell repulsive but that exposure to its vapors caused nausea, a sense of suffocation, and an almost unendurable, long-lasting irritation of the nasal mucous membranes.

Bunsen's discovery of cacodyl, exciting to chemists as the first organic compound containing a metal, was part of a larger mission. Throughout the 1800s, chemists, including Bunsen, worked painstakingly to isolate and thus discover new compounds, particularly new elements. The challenge was how to chemically and physically separate elements and then distinguish a small sample of one silvery metal, or one grayish crystalline powder, from another. This is where Bunsen's burner came in. Nineteenth-century chemists knew that when different pure elements were burned in a flame, they produced distinctive colored flames. Burning sodium produced an orange burst of flame; copper, blue; and zinc, bluish-green. It's these burning elements that produce the fantastic, flaring colors of fireworks. Bunsen reasoned that it would be possible to isolate and discover unknown elements based on their flame color. To do this, he wanted a burner that produced a hot flame but very little light, so that the flame's light wouldn't compete with the light emitted by the chemicals he was burning. Working with his lab technician, Bunsen finessed existing designs of coal gas burners, adding the critical air baffle at the bottom of the combustion tube to create a burner whose flame could be rendered almost invisible. It's this characteristic of the Bunsen burner that makes fine-tuning the flame such an alluring experience—getting the near-invisible flame. Yet, even with a colorless flame, Bunsen found that many elements' flame colors are too similar, as shades of blue-green or red, to be of scientific value. Then the great chemist had an idea. He would hold colored pieces of glass in front of his eye to look for small differences that would let him distinguish between elements. A blue lens would absorb the blue from the glowing element, leaving visible any other perhaps identifiable colors.

In the fall of 1859, while Europe's natural history community was abuzz with the controversy created by Darwin's *On the Origin of Species*, Bunsen, in his stone-walled lab in Heidelberg, set to work with his burner and pieces of colored glass, identifying the colors of light emitted by different chemical "species" (as elements were sometimes called) when they were heated in the burner's flame. This is how Bunsen's friend and colleague physicist Gustav Kirchhoff found Bunsen when he rolled into the chemist's lab one day in the summer of that year. Kirchhoff, using a wheelchair as a result of a childhood accident, had become one of Europe's greatest physicists, accomplishing fundamental work in electricity and later developing the concept of blackbody radiation—the characteristic emission of different wavelengths of light by an object at a specific temperature. The two scientists had met at the University of Breslau, and Bunsen had been instrumental in getting Kirchhoff a position at Heidelberg. It would later be said that Kirchhoff was Bunsen's greatest discovery.

Used to comfortably brainstorming experimental ideas, Kirchhoff immediately pointed out that there was a far easier and more precise way to study an element's emitted light—use a prism to break the light into a spectrum. The idea was far from new, but even in the mid-nineteenth century it held an aspect of the surreal from the time when Sir Isaac Newton, three centuries earlier, had conducted the first modern scientific light experiments. In Newton's day, *spectrum* was synonymous with *specter*—a phantom or apparition. This etymology gives a sense of Newton's awe when, having bought a prism at a local market, he carefully drew closed the curtains in a room until only a single shaft of sunlight pierced the room's gloom. Then the über-savant placed his prism in the light's path. It was a light miracle. On the side of the prism closest to the curtain, a shaft of white light entered it, but when the light exited the prism, it projected on the far wall a rainbow of color.

Many others had already observed that white light actually consisted of a rainbow of colors. What set Newton apart was that he didn't just see the rainbow for itself but that he understood that sunlight held a deeper story, one that could be teased apart by separating the light. He showed that, by using a second prism, the rainbow of light could be reconstituted into white light and that the rainbow effect was created by the different colors of light being refracted, or bent, in different amounts by the prism, thus adding an all-important level of quantification. Newton explained all this in 1672, in the first paper he sent to the Royal Society, in which he used the word *spectrum* in its modern sense. He'd turned a ghost into a matter of science.

Building on Newton's discovery, Bunsen and Kirchhoff quickly set up a prism, and within hours they discovered something extraordinary. Each of the elements they tested had its own spectral pattern—a light fingerprint. When they heated a small sample of sodium in the burner's flame and examined its light through the prism, they didn't see a full rainbow spectrum. Instead, where there would otherwise be a spectrum, it was mostly dark, except for two bright yellow lines. When they burned calcium, they saw at a glance that its spectral signature was characterized by an intense green line and a strong orange one. Strontium, on the other hand, has no green lines, but it has eight lines that are very prominent: six red ones, one orange, and one blue. To extend their experiment and see more detail, Bunsen and Kirchhoff built a small device—the world's first laboratory spectroscope—a viewing tool (the *scope* part) for studying a substance's spectral signature (the *spectro* part). This first spectroscope consisted of a box containing a prism and a small mirror. Light from a burning substance was funneled into the box via a small "telescope," and the mirror onto which the spectrum fell was also observed with a small telescope.

Bunsen and Kirchhoff were stunned by what they saw. Each element's light fingerprint was as unique a marker of that element as is a person's fingerprint of each individual. Even this notion of clear identification of something, or someone, based on seemingly obscure physical characteristics was itself a surprise. (It's a wonderful confluence of events that the first use of a human fingerprint as a means of identification occurred just a year before.) In the scientific paper reporting their findings, the two savants effused that here was a foolproof way to identify an element no matter what compound it was part of, no matter what the temperature of the flame was, and no matter what type of chemical was used to heat it. Sodium's spectral fingerprint is the same, whether it is coming from sodium chloride or sodium fluoride, or whether it is heated using alcohol, hydrogen, or (this being Bunsen) something the authors called "detonating gas." And it got better. Not only could an element be identified by its light fingerprint, but it required only a ridiculously small sample. A quantity of sodium utterly invisible to the naked eye, as little as one part per twenty million of air, revealed its presence, when heated, by its light signature.

Bunsen and Kirchhoff realized they'd hit on something of enormous scientific potential. "If there should be substances that are so sparingly distributed in nature that our present means of analysis fail for their recognition and separation," they wrote, "then we might hope to recognize and to determine many such substances in quantities not reached by our usual means, by the simple observation of their flame spectra." They were confident in saying this because, as is ever the case in the conservative world of science, they'd already done what they were describing. As they later reported, they analyzed the invisible contents of a bottle of mineral water from Durkheim and found a new element whose spectral fingerprint was dominated by a splendid blue line. They called the new element cesium, Latin for "sky blue." It was the first element named not for its outward appearance, use, or place of origin but for its light fingerprint.

MYSTERY OF THE FRAUNHOFER LINES

As Bunsen and Kirchhoff worked to document the light fingerprint of one element after another, a different spectrum emerged into Kirchhoff's mind's eye: not a single element's spectrum but, incredibly, that of the Sun. In a flash of inspiration, Kirchhoff saw that spectroscopy was the key that might unlock one of the greatest mysteries of his day. With a simple step in the laboratory, he could make a great leap from the elements to the stars.

The mysterious solar spectrum that Kirchhoff was contemplating had stumped scientists for half a century. It was the nineteenth-century equivalent of a cosmic crossword puzzle for which no one could determine even the first letter of the first word. The spectrum had been produced by the deeply talented Austrian optician and telescope maker Joseph von Fraunhofer. Fraunhofer's own scientific journey was phoenixlike. The eleventh child of a struggling glassmaker, Fraunhofer was originally apprenticed to a strict master, with no hope of going to school. However, the master's house collapsed in a freak accident, killing several people but leaving the teenage Fraunhofer unhurt in the rubble. His amazing survival was witnessed by the future Bavarian king, who became Fraunhofer's sponsor. Within years, the young man's talent and drive made him a respected optician and entrepreneur.

As such, Fraunhofer tackled a major spectral problem of his time. Lens makers of the day struggled with small flaws in a lens's manufacture that resulted in unwanted spectra at the edges and sometimes gave eyeglass wearers a rainbow-colored view of the world. For eyeglass wearers, this was merely a bother, but for astronomers it was a fundamental problem, distorting their view of the heavens. Fraunhofer had the clever idea of finding a fixed spectrum by which he could calibrate and adjust lenses—and what better light source for this than sunlight itself? But the solar spectrum he produced with his telescope in 1814 was more of a head-scratcher than a solution. Strewn across the solar spectrum were thin, dark lines. Sitting in his darkened observatory, staring through his rudimentary spectroscope, Fraunhofer stared at this odd spectrum that recalled Newton's specter. It was as if some celestial jester had taken a pencil and ruler and drawn thin lines at irregular intervals across the Sun's spectrum, similar to what would appear today as a solar bar code.

Newton hadn't seen these lines because his spectrum was too diffuse—the equivalent of a low-resolution digital picture in which fine details are lost—and an earlier nineteenth-century astronomer, Francis Wollaston, who'd first detected these lines in 1802, had discounted them as the boundaries between spectral colors. Yet in his more detailed solar spectrum, Fraunhofer clearly saw that many of the dark lines occurred *within* individual colors. In the manner of the deeply obsessed and driven scientist throughout the ages, Fraunhofer sat at his telescope and painstakingly recorded the positions of the soon-to-be-called Fraunhofer lines. He eventually detailed almost six hundred of them, labeling the most prominent ones from A to H. He wondered what could cause these strange but consistently present lines that were polluting the Sun's pure spectrum. He even glimpsed them in the light from several stars. He might well have answered the question himself, but in 1826 he died of tuberculosis at the age of thirty-nine.

For more than thirty years, the origin of the Fraunhofer lines hung as the great cold case of nineteenth-century astronomy. Now, with his laboratory evidence, Kirchhoff thought he'd cracked the case. The key to solving the mystery was the fingerprint evidence; in this case, sodium's light fingerprint. When Kirchhoff saw sodium's spectrum with its dominant yellow lines, he was struck by the fact that he'd seen lines in exactly the same place in another spectrum—the Sun's. The difference was that in the Sun's spectrum, the lines were the dark "D" Fraunhofer lines—two lines, both in exactly the same place in the yellow part of the spectrum, but one bright, the other dark. Was it possible that both the dark and bright lines were produced by sodium?

Kirchhoff performed an ingenious series of experiments that culminated with this one: he modeled the Sun and its possible atmosphere in his lab. From his wheelchair, the physicist heated a piece of charcoal, as a stand-in for the Sun, and then shone its light through sodium gas before recording the spectrum. When he examined the spectrum, he became the first human to know part of what the Sun is really made of—there were the two dark Fraunhofer lines, corresponding to sodium's yellow lines. The Sun's atmosphere, Kirchhoff realized, is salty.

These experiments were the foundation for what is still taught as Kirchhoff's law of emission and absorption. According to Kirchhoff's law, the position of either dark or bright lines in a spectrum is an element's telltale signature. If the line is dark, the element is absorbing the light from another source; for example, elements in the Sun's atmosphere. If it's bright, the element is giving off light as a result of absorbing energy, for example, that from a flame. The relative strength of the lines depends on the density and abundance of the elements. But the kicker, concluded Kirchhoff, was that the chemical analysis of the Sun's atmosphere required only the search for those substances that produce the bright lines that coincide with the dark lines of the solar spectrum. Kirchhoff meticulously drew the solar spectrum with its Fraunhofer lines and placed this side by side with the spectra from thirty different elements. Like a detective identifying multiple fingerprints at a crime scene, Kirchhoff identified the elements present in the Sun's atmosphere. There were a remarkable sixty exact overlaps between the Fraunhofer lines and the laboratory spectrum of iron. The Sun's atmosphere was ablaze with iron, as well as being a chemist's cabinet of Earthly elements, including sodium, calcium, nickel, and magnesium, along with barium, copper, cobalt, and zinc.

Kirchhoff's observation of that first element, sodium, in the Sun's spectrum was as momentous as Galileo's observation of the moons of Jupiter. Galileo knew in a moment that the Sun was not the absolute center of creation—heavenly bodies orbited the planets as well. Kirchhoff realized that the power of the elements' light fingerprints amazingly extended beyond the Earth to the stars. He and Bunsen had developed a tool, spectroscopy, that opened the heavens to a new kind of scientific exploration. "Spectrum analysis . . . opens to chemical research a hitherto completely closed region extending far beyond the limits of the earth and even of the solar system," he wrote.

Auguste Comte was wrong. Kirchhoff knew that it would be possible to lift the Sun and the brighter stars out of a veil of mystery and to determine not just their movements but also exactly what they were made of. In 1861, after a demonstration of this revolutionary spectroscopic technology, Warren De la Rue, a British chemist and astronomer famous for his pioneering photographs of the Sun and Moon, exclaimed to an audience of scientists in London—as if rebuffing Comte—that stellar spectroscopy was akin to having a piece of a star in your hands. "If we were to go to the Sun, and to bring some portions of it and analyze them in our laboratories, we could not examine them more accurately than we can by this new mode of spectrum analysis."

Kirchhoff and Bunsen thus created the most powerful, and most unsung, astronomical tool after the telescope. Stargazers previously had pitifully little to work with, compared with their terrestrial scientific colleagues in chemistry or biology. Astronomers could describe a star's position, comparative brightness, and observed color, but not much else beyond what can be seen by looking up into the night sky. Stellar spectroscopy presented a previously unimaginable way of knowing a star, making a point of light something physical and elementally related to the Earth and to us. Telescopes capture light, but it is spectroscopy that makes this light meaningful.

It's difficult to underestimate the impact of stellar spectroscopy on astronomy. The vast majority of what we know about stars—their composition, temperature, and movement—is due to spectroscopy. It enables us to tease out the incredible information starlight carries. This light message can travel from a distant star, across the known universe, without losing a single word of the story it carries. For astronomers, it was as if for decades they'd been receiving messages by mail, with the only information about the sender being a return address and the stamp on the envelope. Suddenly astronomers realized that they could open the envelope and find inside a letter containing an amazing story about the stars.

ORDER IN THE HEAVENS

Astronomers soon realized that stellar spectroscopy revealed more than just what stars were made of; it revealed how they were related. Just as the Victorian zeitgeist for categorizing inspired geologists to classify rock types, and natural historians to minutely describe and classify newly discovered species of animals and plants, astronomers now had a previously unimaginably powerful tool for bringing order to the heavens. Indeed, the discovery of stellar spectroscopy was greeted as a gift from the heavens by the Englishman William Huggins. In 1856, he'd sold the family's silk business—in the face of competition from the new phenomenon of department stores—and moved from London to Upper Tulse Hill on the city's southern outskirts where, under a dark night sky without London's smog or light pollution, he oversaw the building of a world-class observatory.

Although Huggins was nominally an amateur gentleman astronomer, he was ambitious and was looking for a challenge beyond the then run-of-the-mill observing of the positions and motions of stars. He found it in stellar spectroscopy. "The news reached me of Kirchhoff's great discovery of the true nature and chemical composition of the sun from his interpretation of the Fraunhofer lines," Huggins later recounted. "This news was to me like the coming upon a spring of water in a dry and thirsty land. Here at last presented itself the very order of work for which in an indefinite way I was looking—namely, to extend his novel methods of research upon the Sun to the other heavenly bodies."

Aided by his neighbor, William Miller, a professor of chemistry at King's College, Cambridge, and an early champion of laboratory spectroscopy, Huggins set to work, laboriously comparing stellar spectra with those of Earthly elements. In this way, Huggins and Miller studied fifty stars and compared their spectra with those of twenty-seven elements. To their amazement, no matter where they looked in the heavens, they saw evidence of the same elements. For Huggins, it was a profound realization: the stars are distant but are not fundamentally different from us. It was this connection, rather than the relationship between the stars, that fascinated Huggins when he and Miller published their findings with the Royal Society in 1864, marking the start of stellar astrophysics. They reached a stunning conclusion: not only was it possible to know the composition of the stars, but, based on this analysis, they concluded that there existed a common chemistry throughout the cosmos. A commonality others would soon see could be used to make cosmic comparisons.

For the comparison of stellar spectra to really take off, the intersection of astronomy with another transformational new way of seeing was required. On a clear night in 1872, American astronomer Henry Draper was in his observatory at Hastings-on-Hudson, about twenty-five miles north of Lower Manhattan. That evening, rather than simply observe the spectrum of the bright star Vega, Draper took a photograph of it. Although this might not seem like such a leap, it was the marriage of two world-changing technologies—spectroscopy and photography. Draper's first spectrogram—a visual record of a star's spectral fingerprint—meant that rather than sit in the dark and carefully record by hand a star's spectrum, astronomers could take a picture of it and then examine it at leisure in daylight. Imagine the delight of a police booking officer who had previously been required to hand-draw each suspect's fingerprint, but who was now able to use an ink pad to quickly and accurately record the print for later use. For astronomers, it was even better than this. Before spectrographs, they could see the spectra of only the very brightest stars—those bright enough to create a spectrum visible to their eyes. Now, with an exposure time of minutes or hours, astronomers could patiently collect photons on photographic plates from thousands, and soon hundreds of thousands, of faint and faraway stars not visible to the naked eye. Suddenly, a vast and literally endless bounty of stars and other celestial objects were made visible to spectroscopic fingerprinting.

Following Draper's death at the age of forty-five in 1882, his widow, Anna Palmer Draper, also an accomplished astronomer who'd observed alongside her husband, wanted to commemorate her late husband's stellar love. She found the perfect man to carry on his vision in Edward C. Pickering, the director of the Harvard College Observatory in Cambridge, Massachusetts. With seed money from Anna Palmer Draper, Pickering created the Henry Draper Memorial, one of the great star-cataloging missions of all time. Pickering saw in the combination of a well-staffed observatory and spectrographs the opportunity to collect thousands of stellar spectra that would allow him to discover stellar relationships and to organize them based on their common characteristics.

During a thirty-five-year period, Pickering led a series of spectroscopic star surveys and cataloging efforts that by 1924 listed the spectra of more than 225,000 Northern and Southern Hemisphere stars—organized in a nine-volume series called the Henry Draper Catalogue—or HD Catalogue. Fueled by Pickering's vision and fundraising, his assistants did the vast bulk of the work: the observing, the spectral comparison, and the analysis that led to the HD Catalogue's historic conclusions. In a 1913 photograph, Pickering stands suited in the back row of his "harem": thirteen middle-aged women wearing ankle-length dresses and blouses with high-neck collars and with their hair up in neat, controlled buns. These were the "computers"—the astronomy field's version of secretarial pools—women hired to do the detailed analysis of the astronomical plates, as well as related mathematical calculations. For example, Annie Jump Cannon, an astronomer educated at Wellesley and Radcliffe Colleges and hired as assistant astronomer, led the classifying of almost a quarter-million spectra. In the annals of astronomy, Cannon is the greatest star fingerprinter of all time: between 1896 and her death in 1941, she classified the spectra of an estimated 395,000 stars.

In 1922, this herculean effort led to the International Astronomical Union adopting a stellar classification system based on the dominant elements in a star's spectra. Thus was born the mnemonic device remembered by undergraduate astronomy students ever since: *Oh Be A Fine Girl/Guy, Kiss Me!* This was based on the seven main stellar types: O, B, A, F, G, K, and M, arranged in order of decreasing stellar mass and surface temperature, from the biggest, hottest blue O-stars on the left, to the smallest, coolest red M-type stars on the far right.

Less than a century after Auguste Comte's prediction that the stars would forever remain unknowable, astronomers had turned light's secret code into a way not just of knowing the composition of stars but also of seeing underlying order in the heavens. This new cosmic code was telling a deeper story. Comte's seemingly irrefutable logic had been transformed by the combination of spectroscopy, telescopes, and photography. A new breed of astrophysicists could indeed experiment by analogy with stars in their laboratories by observing the spectra of elements and then looking for them in the stars. And though astronomers still knew of only one solar system in the cosmos, it was now possible to compare the chemical structure of different corners of creation.

Yet in all this stellar and terrestrial elemental fingerprinting success, there was a puzzling hole right in the heart of the great ordering system of the elements, the periodic table. One element remained elusive, a blank space in the common cosmic chemistry: element number 43. Where we'd find this mysterious element would change not only our understanding of the elements and the stars but also of ourselves.

A STRANGER IN THE STARS

Mount Wilson astronomer Paul Merrill was an unlikely Moses of the Stardust Revolution, bringing down from a mountain a key clue in 1952 that would point the way to the elemental laws of our stellar origins. The thing was, Merrill wasn't looking to change the world. He liked it well enough the way it was. The son of a minister of the Congregational Church, he was active in his Pasadena, California, church community and was a staunch Republican. Daily at lunchtime, from the 1930s to 1950s, he'd leave his Santa Barbara Street office in the white-stuccoed Carnegie Institute and walk past the neighboring arts-and-crafts-style bungalows for the Reyn Restaurant, still located around the corner on Lake Street. Before entering, if the day was clear, he could look up to see Mount Wilson rising into the blue California sky, and then he'd head for his table and its marbled-blue vinyl bench. If it was an odd-numbered year, the waitress would bring a peanut butter sandwich. During even-numbered years, she'd deliver a fried egg sandwich, in which case his tablemate, friend and fellow Mount Wilson astronomer Alfred Joy, would have the peanut butter sandwich.

Merrill's penchant for consistency and detail found its deepest expression in his work. He was, one long-time colleague observed, "a most seriously, serious scientist." On many afternoons, he'd head up the winding road to Mount Wilson. Many nights, he'd see Edwin Hubble seated at the base of the huge one-hundred-inch-diameter Hooker telescope, imaging distant galaxies as part of his effort to calculate their speeds—an effort that would underpin the big-bang theory. There were no niceties exchanged between the two, such as "Hey, Ed, what's the latest red shift?" Merrill detested Hubble, and the feeling was mutual. Merrill could "cloud over like a summer thunderstorm blackening the Minnesota plains" at the mere mention of Hubble's name. Merrill thought Hubble was a scientific showman, courting media sensationalism around his expanding-universe discovery, and just plain wrong when it came to some of his measurements of the distances to galaxies and the brightness of distant stars.

Fueling this irritation was the fact that Merrill felt underappreciated. He was a spectroscopist, doing what he felt was the most important aspect of modern astronomy—work largely ignored by the media in favor of Hubble's 1929 discovery of the possible expansion of the universe, which had riveted the public imagination like no previous story in science. To add insult to injury, Hubble's measurements of "red shift" in fact depended on spectroscopy. The red shift of light is measured by comparing the location of the Sun's Fraunhofer lines with those of distant objects. If the lines are moved toward the red end of the spectrum, the light is being red shifted.

For all their differences, Merrill and Hubble were both the type of men George Ellery Hale wanted for his observatory. He realized that some astronomers are what their peers derisively call stamp collectors—focused on gathering, sorting, and describing cosmic objects—whereas others are intent on putting the evidence together into a larger story. Hale believed that the key was to combine both types of personalities and the best possible instruments, which he did at Mount Wilson. On the hundred-inch Hooker telescope, Hale installed a spectrometer with unprecedented resolution, inspired by Bunsen and Kirchhoff's insights. In 1919, he hired Merrill, who already as a young man had demonstrated he was a master at using spectroscopy to tease apart light in order to determine what stars are made of, how they're changing, and how they're moving.

During his forty-three-year official career on the staff of the Mount Wilson Observatory, Merrill accumulated what one colleague described as "a truly monumental mass of carefully compiled observational material. . . . A gold mine of factual information for those who must eventually construct sound physical theories to account for their properties." In effect, like the emerging paparazzi in the valley below him, Merrill took countless spectrograms of the stars with the hope of being able to gain some understanding of their lives.

If in all else Merrill was a company man, when it came to the stellar company he kept, the astronomer liked the oddballs. He had what a colleague described as a "passionate enthusiasm for the study of those stars whose spectra show deviation from the normal," the so-called variable S-stars—a parallel subtype of M-stars. Since their discovery, Mira variable stars—named after the largest and brightest variable S-star, Mira—had intrigued astronomers, most notably because they challenged the long-held belief in the unchangeability of stars. Not only do these bright red stars change, growing alternately brighter and then fading as if they were on a cosmic dimmer switch, they do so in a characteristic fashion. Merrill was particularly interested in long-variable stars, those that pulsated over a period of more than one hundred days—such as R Geminorum, a long-period red variable star located between the stick figures in the Gemini constellation. We now know that these S-stars are Sun-like stars in their death throes, having bloated to many times their original size. Their atmospheres periodically pulse in and out, in the star equivalent of menopausal hot flashes.

By early 1952, Merrill was just months from retirement. His wrist joints were so swollen from years of severe arthritis that he had to use both hands just to hold a cup of coffee. Worst of all, he could no longer endure the cold and the physical effort required for observing at the telescope. Then he received a series of eight spectrograms like none that he, or anyone else, had ever seen. The images were some of the first taken using the spectrograph on the new, massive two-hundred-inch Hale Telescope on Mount Palomar, one hundred miles to the southeast. For any astronomer, a bigger telescope is the difference between seeing and not seeing, between knowing and not knowing; and all astronomers knew that the Hale Telescope was arriving as a conquering giant. Waiting for it to come into full operation was like waiting for Christmas morning. During the testing and adjusting of this new world record–holder, Merrill convinced his Carnegie colleague Ira Bowen to take the spectra of eight bright S-type stars. "I am greatly indebted to him for these valuable photographs," Merrill wrote in the scientific paper that reported his findings.

In his office on Santa Barbara Street, Merrill hunched over the treasured photographs and inspected the complex forest of dark and bright lines that made up the spectrogram. It's not known in which of the eight spectrograms he first saw it—maybe the spectra of R Geminorum; for five hours and five minutes on March 23, 1951, the Hale Telescope had focused on this red giant star in the middle of the Gemini constellation, and the Hale Telescope's spectrograph had captured and sorted thousands of the star's photons. As the world's highest-resolution stellar spectrogram, it was an extraordinarily detailed and complicated spectra—a proverbial forest in which it was difficult to see the trees. Merrill spotted the usual S-star suspects. There were the characteristic absorption lines for zirconium oxide and titanium oxide. There were the well-known emission lines (S-stars are different from most in also having emission lines from excited atoms) for the elements iron, manganese, and silicon. But amid the light fingerprints of these familiar elements was a stranger—absorption lines that Merrill, in almost a half a century of looking, had never seen. Eventually he nailed down the identity of this mysterious stellar element: technetium. It had no business being there, yet technetium would change our view of the cosmos, rather than the other way around. After all, Paul Merrill was a stickler for details, and starlight doesn't lie.

Today, technetium is well known as the primary radioactive element used in nuclear medicine. If you've ever had a heart stress test or breast lymph node imaging, you've been injected with a tiny amount of this radioactive element. Worldwide, it's injected into millions of patients a year, but until 1937 it was a mystery. When the Russian chemist Dmitri Mendeleev mapped out the modern version of the periodic table, this ordering system of columns and rows included blank squares—elements predicted by the table's structure but not yet discovered. Almost dead-center on the periodic table, between molybdenum and ruthenium, and kitty-corner to iron and chromium, was a missing tooth in the periodic table's otherwise solid, logical smile. For more than a century, element 43 had evaded detection.

Chemists tried and failed to isolate element 43 from hundreds of minerals. Technetium was so elusive because it is a most unusual element. In 1937, the Italian chemists Emilio Segrè and Carlo Perrier received a parcel in the mail that contained a rare, exotic, and infinitesimal—less than a billionth of a gram—sample of unknown material. Sent by a colleague at the University of California–Berkeley, the material had been produced by a new machine of the nuclear age, a cyclotron. It enabled scientists to smash together atoms at speeds at which nuclear reactions take place. The tiny amount of silvery metal sent by mail had been produced by bombarding molybdenum with deuterium, a variant of hydrogen that contains an additional neutron. But when Segrè and Perrier analyzed their sample, they did not find element 42, molybdenum; instead they became the first to glimpse element 43. And they'd found not only the elusive element 43 but also the first "Frankenstein element"—the first element created by nuclear alchemy. It wasn't until 1949 that they named it technetium, from the Greek *tecknetos*, for "artificial." The next year, technetium's light fingerprint was thoroughly documented by scientists at the US National Bureau of Standards.

It was readily apparent why technetium was so difficult to find on Earth. Technetium is the elemental equivalent of an ephemeral flower bloom. Alone among the elements at the heart of the periodic table, it has no stable, nonradioactive form. Today we know of nineteen isotopes, or elemental variants, of technetium, but the majority decay into other elements in days or hours (the reason it's an ideal tool in nuclear medicine, since it readily disappears from the body), while the longest-lasting isotope has a half-life of about four million years. Compare this with uranium, the half-life of which is 4.47 billion years, about the age of the Earth. This means that when technetium is produced, even the longest-lasting isotopes will have disappeared beyond detection in less than ten million years.

In April 1952, Merrill traveled to Washington, DC, for the annual meeting of the National Academy of Sciences, at which he presented a paper titled "Technetium in the Stars." He ended his remarks by offering three possible ways to interpret his surprising findings. The first two possibilities seemed unlikely: perhaps a stable isotope of technetium exists in stars, one not yet found on Earth—though Merrill knew that the National Bureau of Standards scientists suggested this wasn't the case; or some stars are so young as to still contain detectable levels of technetium, which still required a longer-lived version of technetium. The third possibility, concluded Merrill almost reluctantly, was that "S-stars somehow produce technetium as they go along," that while he stood there presenting, stars were actually making technetium. After all, what was possible in cyclotrons at Berkeley was perhaps also possible in the stars. Merrill didn't speculate any further. He didn't need to. His audience would have seen the inference: if stars make technetium, an element in the heart of the periodic table, they aren't just balls of fire but factories of the elements. Without intending it, and while never embracing the implications of his historic discovery, the man who eschewed big talk about an expanding universe discovered the key to an evolving one. He set a spark that helped fuel one of the greatest struggles of the Stardust Revolution: the quest for the origin of the elements.

CHAPTER 3
THE ORIGIN OF THE ELEMENTS

It is a rather interesting coincidence that physically a man is nearly a mean proportional between an atom and a star. It requires about 10^{27} atoms to make a human body and the material of 10^{28} human bodies to make an average star.

—Walter Adams, director, Mount Wilson Observatory, 1928

OF STARS AND ATOMS

Two miles down Pasadena's palm-lined roads from the Carnegie Institute and Paul Merrill's former office is a rectangular intellectual enclave that is the other half of George Ellery Hale's revolutionary vision: the California Institute of Technology. Among the squat buildings on the southern edge of Caltech's leafy campus is a windowless, five-story stuccoed edifice, the W. K. Kellogg Radiation Lab. In the midst of the Great Depression, the Caltech Nobel laureate Robert Millikan—the man who'd weighed the ephemeral electron—used his prestige to bend the ear of William K. Kellogg, the great breakfast cereal inventor, industrialist, and philanthropist. Millikan convinced Kellogg to support a new kind of lab at Caltech, one with extremely high-powered machines that would use light and particles to probe the essence of atoms—and that might also be able to treat cancer patients by using high-energy x-rays to kill tumors. Kellogg bit at the idea, in the process unwittingly joining together the realms of breakfast cereal and the deepest secrets of the cosmos.

The Kellogg lab's windowless nature spoke to Hale's vision for Caltech. This lab wasn't like the observatories up the mountain, which looked out. This one looked within. To pierce the mysteries of the biggest objects in the universe—stars—Hale believed it would be necessary to know the nature of the smallest objects—atoms. This was the new age of astrophysics, one that Hale believed was a two-faced chimera: it could be understood as either astronomy or physics, at once immense and minuscule, and both out there and the essence of here.

Entering the lab today through its heavy, rust-edged main door, you find yourself in a world parsed down for high-energy physics research. The hallway walls are spartan cement, broken up periodically by the original wooden doors of professors' offices. Farther along the hallway, past the chained-off doorway with the sign warning "Radiation Hazard," is the low-ceilinged, bunker-like common area. The room's worn carpet is dotted with tables and a sofa, and the walls are lined with old scientific journals and books. It feels like a museum as much as a workplace. There's the sense of physicists' ghosts chatting intently, still arguing about the results of a particle accelerator experiment from the lab's heyday in the 1950s, when the nuclear physicists here led the world in probing the nature of atoms. Hanging over the kitchenette sink is a framed photo commemorating that glorious time. The man in the black-and-white photo is William Fowler, the physicist whose name is now most intimately linked to the Kellogg lab.

With his neatly trimmed white beard and jovial energy, Fowler would have made a great Santa Claus. In the photo, flanked by a cheering crowd of fellow Caltech staff, students, and spouses, he's holding up a sports jersey emblazoned with the words "Nuclear Alchemist" over a large number "1." It's October 1983, and Fowler has just been awarded the Nobel Prize in Physics for, as the Nobel committee put it, "his theoretical and experimental studies of the nuclear reactions of importance in the formation of the chemical elements in the universe." Or, more prosaically, for his role in unraveling one of the greatest natural mysteries of all time: the origin of the elements, the way that stars forge the cosmic alphabet of atoms, from argon to zirconium.

What the photo doesn't reveal is that, for all his public exuberance, Fowler was haunted by the recognition from Stockholm. It had forced him to choose between two great loves. On the one hand was "the lab," his home and family for almost half a century, where he'd started as a graduate student and was now the venerated patriarch and administrative defender. On the other existed a singular, personal, and scientific friendship—and the truth that he knew it held. To get a glimpse behind Fowler's smiling facade and into the complexity of what he was feeling, you have to walk across the Caltech campus and don cotton gloves in the university's Ivy League–quality archives. There, among the thousands of letters, lecture notes, scientific jottings, and articles that Fowler meticulously kept—including correspondence with a who's who of twentieth-century physicists—is a photocopy of an airmail letter. It's dated November 3, 1983, and is addressed to his friend of thirty years, the renowned British astrophysicist Fred Hoyle.

They'd met in Fowler's Kellogg lab office in 1952, when the brash young Brit, on a visit to Mount Wilson, tracked down Fowler with a preposterous suggestion. Hoyle proposed that carbon, the sixth element in the periodic table and the cornerstone of life, was made in stars. Not only that, but he was sure that Fowler's lab had the atomic tools and the know-how to prove Hoyle's hypothesis. Fowler's colleagues in the room at the time dismissed Hoyle's hypothesis as the rantings of an astronomer way out of his league in atomic physics. After all, by this time the Kellogg lab physicists were second to none in measuring atomic transmutations. Their quantum mechanical calculations showed that there was no way carbon could be forged in stars. Impossible. Nevertheless, whether because of Hoyle's intensity and evident intelligence or due to Fowler's enormous gregariousness and willingness to step into the unknown, Fowler didn't escort Hoyle to the Kellogg lab's heavy steel-doored exit. Instead, he gathered colleagues and started planning a momentous experiment.

Fowler may have recalled this meeting as he sat in the office where the two men had first met and where he wrote to Hoyle after receiving the call from Stockholm:

Figure 3.1. Feted by his Caltech colleagues, new Nobel laureate William Fowler's smile belies his distress over the fact that the #1 jersey in fact belongs as much to Nobel-less Fred Hoyle. *Photo reproduced with permission from the Archives, California Institute of Technology.*

> *Dear Fred,*
>
> *After the initial elation and excitement I have had a heavy heart for two weeks. It is impossible to understand why the prize was not given to you or shared between us. I realize that nothing I can write will help but this personal note to you helps relieve my own feeling of hurt.*

But, Fowler continued, for the sake of the Kellogg lab (which, though he didn't mention it, was financially in dire straits) and his colleagues there, he couldn't turn down the world's highest scientific honor. As if in a Shakespearean tragedy, one of the greatest scientific mysteries of all time ended with star-crossed friends facing choices that could never be fully satisfied. Together they'd shared a cosmic quest for the origin of the elements. It was an intellectual journey that would take them from the beginning of time and into the hearts of stars. And it would reveal that the three greatest secrets of the universe are in fact strands in a single cosmic cord: what makes stars shine, the origin of the elements, and the birth of all we see. These were cosmic riddles that for three hundred years perplexed and at times tortured the greatest minds in science. They began with the alchemist's dream to turn base metals into gold and thus to know the mind of God.

THE ALCHEMIST'S DREAM

For centuries, the origin of the elements was the elephant in the room of scientific questions: Where does all this stuff ultimately come from, me, you, the metals that make your watch, the gold in your wedding ring, the air we breathe? And, just as importantly, can one element be changed into another? This last question was the basis of the alchemists' Philosopher's Stone—the knowledge of the power or substance that could transform base matter into precious gold. Medieval and Renaissance alchemists pursued an ancient belief that there existed some primordial matter or spiritual essence that directed matter to be one element or another and that, with enough work and purity of spirit, it would be possible to grasp this timeless, all-powerful knowledge—to wield the Philosopher's Stone.

As such, modern science has its roots in the quest for the origin of the elements. You probably know Sir Isaac Newton from high school for his laws of motion and optics, but more often than not his prodigious mind raced with images of lead, gold, suns, moons, serpents, and dragon's wings. At heart, Newton was an alchemist, probably the greatest ever. The father of modern science spent more time in his private shed attached to Cambridge University's Trinity College trying to secretly transmute elements than in working on his monumental theories of gravity or light. His alchemical experiments lasted twenty-five years, during which he carefully amassed notes and reflections that could have formed the equivalent of a five-hundred-thousand-page book on the topic. This painstaking effort wasn't in quest of riches but was in search of a far greater jewel: to share in the act of creation, to know the mind of God. "Just as the world was created from dark Chaos," Newton reflected in a note written in the 1680s, "through the bringing forth of the light and through the separation of the fiery firmament and of the waters from the earth, so our work brings forth the beginning out of black chaos and its first matter through the separation of the elements and the illumination of matter."

Yet, after a quarter century of secretive effort, a despairing Newton shelved his crucibles and vials of mercury and abandoned his search for the origin of the elements. The world's greatest scientist was stumped. Little did he imagine that he'd paved the way: his understanding of gravity and light would eventually illuminate the way to understanding the origin of the elements. In the realm of scientific timing, Newton was three hundred years too early.

By the late 1800s, talk in the physical sciences about the origin of the elements was considered to be in bad taste. Although biology had entered the new world of Darwinism and evolutionary change, astronomy, physics, and chemistry were still largely grounded in a visceral and intellectual sense of permanency. Just as nature abhors a vacuum, science abhors an untestable hypothesis, and the origin of the elements appeared to live squarely in the land of speculation. However, the Victorian-era British polyvalent Sir William Crookes wasn't a man to shy away from an unpleasant topic. He'd been knighted by Queen Victoria for his contributions to solving one of Britain's most pungent and deadly problems of the era, urban sewage disposal. Crookes had penned titles such as *A Solution of the Sewage Question* and its follow-up, *The Profitable Disposal of Sewage*. He'd survived the discovery of what would become a favorite assassin's poison: a leading chemist, he'd discovered the toxic element thallium and named it from the Greek *thallos*, for "green shoot" or "twig," after its telltale bright-green Bunsenian light fingerprint. For Crookes, identifying an element from the hurly-burly of all other elements on Earth only further piqued his rich curiosity and brought him to the nagging question of what natural process produced the element. That was the topic of his presentation on the evening of February 18, 1887, when London's scientific elite arrived by carriage and foot at the British Association for the Advancement of Science's Royal Institution to hear his lecture, "Genesis of the Elements."

"In the very words selected to denote the subject I have the honour of bringing before you, I have raised a question which may be regarded as heretical," Crookes began, warming up his audience.

> At the time when our modern conception of chemistry first dawned upon the scientific mind, the average chemist . . . regarded his elements as absolutely simple, incapable of transmutation or decomposition, each a kind of barrier behind which we could not penetrate. If closely pressed he said they were self-existent from all eternity, or that they had been individually created just as we now find them at the present day. Or he might argue that the origin of the elements did not in the least concern us, and was, indeed, a question lying outside the boundaries of science.

In mentioning eternal elements, Crookes was referring to the work, some decades earlier, of his countryman, the Manchester schoolteacher and chemist John Dalton. In 1805, Dalton had published a remarkable list in the *Memoirs of the Literary and Philosophical Society of Manchester*, but there was nothing literary about the list: the story Dalton told was all about numbers and the elements. He'd resurrected the ancient Greek idea of the atom—that all matter was composed of tiny, indestructible particles. Distancing himself from the alchemists, Dalton argued that each element had its own unchanging atom, thereby barring any alchemical transmutation.

The evidence for this, he said, was in the way elements combine to form compounds. Experimenting with combining different elements, he'd discovered that they always combined in fixed ratios and that compounds contain specific amounts of each constituent element. The most well-known modern combination is water, H_2O, in which every molecule of water is a marriage of two hydrogen atoms and one oxygen atom. On the basis of these fixed mixing ratios, Dalton could identify the intrinsic atomic weight of the then-known elements; that is, how much each element weighed in relation to any other. He assigned hydrogen, the lightest element, the number 1. Using Dalton's system, carbon had atomic weight of about 4, while sulfur weighed in at about 14.4. Thus he showed that the elements didn't just have different chemical properties but, based on their mass, appeared to be completely different atoms.

Crookes knew there was another way of interpreting Dalton's observations. Not surprisingly, while Dalton insisted that his atomic-weight tables showed that the elements weren't transmutable, others saw in Dalton's table just the opposite story. One of them was the English physician William Prout in the early 1800s. When he wasn't lancing boils, treating gout, or setting broken bones, Prout, like Dalton, was fascinated by the burgeoning new field of chemistry. When he looked at Dalton's atomic-weight tables, he saw a pattern of wonderful Pythagorean beauty: each of the elements had a weight that was essentially a multiple of 1, or hydrogen. Perhaps, Prout conjectured, in what's become known as Prout's hypothesis, all the elements somehow begin with hydrogen. "If the views we have ventured to advance be correct, we may also consider that protyle of the ancients to be realized in hydrogen." The "protyle" he referred to was from Greek philosophers' *proto-hyle*, or "first stuff," which some thought was the essence of all matter.

Over the next half century, Prout's hypothesis fell in and out of fashion. The biggest knock against it was that not all the atomic weights were exact multiples of hydrogen, and some were quite far off. No one at the time knew of isotopes—varieties of an element with the same number of protons and electrons but with different numbers of neutrons, giving them all similar chemical characteristics but different atomic weights. For some theorists, Prout's hypothesis was boosted by Mendeleev's periodic table. The genius of the Russian chemist's ordering system is that it is primarily based on atomic weights. Hydrogen, the lightest element, is number 1 and is located in the upper-left-hand corner of the periodic table, while much heavier lead is located down toward the bottom-right corner. Some chemists looked at this sequential buildup of atomic weights and saw deeper meaning. The seemingly intractable alchemist's dilemma remained: If hydrogen was the alpha element, how did it turn into all the others?

The Victorian-era man of science William Crookes had an answer. "In these our times of restless inquiry," he reflected, "we cannot help asking what are these elements, whence do they come, what is their signification? . . . I venture to say that our commonly received elements are not simple and primordial, that they have not arisen by chance or have not been created in a desultory and mechanical manner but have evolved from simple matters—or indeed from one sole kind of matter." What he offered next really helped him to make his mark. Crookes made the first systematic proposal that the origin of the elements had not been found because researchers had been looking in the wrong place. The origin of the elements isn't the Earth; the elements, he offered, are forged in stars.

Minutes later, he held up a small glass jar to demonstrate the technical prowess that undergirded his science. It was an early vacuum tube—a glass tube from which air had been mechanically removed. "I have in this glass tube," he continued, "perhaps the nearest approach to perfect emptiness yet artificially obtained." He explained that his little jar, with a volume of five cubic centimeters, had been exhausted to just one part in fifty million of the air in that room. He described, to his by-now rapt audience, how his experiments involved zapping small samples of metals placed in similar vacuum tubes with high-voltage electrical discharges and studying the spectrum of the light they emitted.

Crookes reminded his audience that it was part of the chemist's creed that each element has a distinctive spectral fingerprint, such as thallium's green emission line and sodium's bright yellow band. Yet when he examined the spectra of the substances in his electrical discharge tubes, he found that the spectra varied. The reason, he surmised, was that the intense energy of the discharge tube rendered the seemingly immutable elements into even smaller bits. (Crookes was right: he'd created plasmas, atoms stripped of electrons. But the electron itself wouldn't be discovered until 1897, a decade later, by J. J. Thomson and his colleagues at Cambridge University's Cavendish Laboratory.) Crookes argued that in the intense heat and electrical forces present in the Sun, matter was reduced to Prout's protyle, or first stuff. He explained that as this plasma cooled, the various elements were formed or frozen out—the lightest first, followed by those with heavier atomic weight. In one of the most wonderfully coincidental intersections in science, Crookes described his stellar ladder of the elements as a double helix, with hydrogen at the top of one strand, moving down to uranium. There was, he concluded, "a genetic relation among the elements."

A RECIPE FOR SUNSHINE

There was a singular, aggravating problem with Crookes's proposal that the elements were formed in the Sun and other stars. No one really knew what the Sun *was*. More precisely, no one knew what made it shine. The search for the origin of the elements had run full-on into one of the other great scientific mysteries of the day. For all of human history, the Sun had risen in the east and set in the west, and on each daily journey it glared like a burning question mark in the sky. The search for the ingredients of sunshine stretched over almost a century and involved many of the greatest scientists of the period.

As a scientific problem, the Sun presented two main challenges: the enormous energy it produces and its age. For more than a century, physicists speculated about what substance could produce enough heat to keep sunbathers tanned for millennia. The first candidate was coal. But in 1848, the German physician Julius Robert Mayer calculated that a Sun-sized chunk of coal, a sphere about a million miles across, would burn for about five thousand years. Even by biblical standards, that wasn't long enough to have warmed Adam and Eve and not extinguished in a sooty poof by the start of Queen Victoria's reign.

Here was the second part of the Sun enigma. The Sun didn't just pump out vast amounts of heat and light; it had been doing so for a long time. Mayer and others realized that the Sun must be powered by something other than chemical energy, and the next contender was something Newton had also thought about—gravity. The champion of a Sun heated by gravity was Lord Kelvin, now perhaps best known for giving the world the concept of "absolute zero" on his eponymous Kelvin temperature scale. Lord Kelvin and his German colleague Hermann von Helmholtz proposed that the Sun's energy came from the gradual gravitational contraction of gas. On one level, this makes perfect sense: if stellar matter was continually squished against itself, as it surely was, this friction would generate an enormous amount of heat. If the Sun was formed from a cloud of continually contracting cosmic gas, Kelvin estimated its age at about thirty million years.

Kelvin and von Helmholtz's model, however, still ran into the age problem. Nineteenth-century geologists had realized that the Earth is hundreds of millions, if not billions, of years old. Here was the rub: it didn't make sense that the Earth was older than the Sun, as calculated from Kelvin's gravitational-heating model. Kelvin, a fundamentalist Christian who never accepted Darwinian evolution, stuck to his gravitational-heating model, preferring his faith to geological evidence. As a result, he concluded, humanity's future looked dim. If the Sun shone by gravitational contraction alone, sooner or later it would shrink to such an extent that the Earth would face the ultimate energy crisis: its solar light and heat would shut off. "We may say, with certainty," Kelvin said in 1862, "that the inhabitants of the

Earth cannot enjoy the light and heat essential to their life for many millions of years longer, unless sources now unknown to us are prepared in the great storehouse of creation."

Fortunately for all involved, as Kelvin argued for a gravity-heated Sun, a previously unimaginably powerful source of energy came to light: radioactivity. The French physicist Pierre Curie showed that the radioactive element radium possessed energy like no other known substance. Pound for pound, radium generated a million times as much energy as dynamite, and not only was radioactivity powerful, it was long-lived. At Cambridge, the atomic physicist Ernest Rutherford realized that radioactive substances have clear ages, measured in half-lives—they emit energetic particles, or decay, in predictable intervals. Radon, the first element Rutherford examined, has a half-life of just less than four days for its most stable isotope. But uranium, he determined, has a half-life of 4.5 billion years. Here, finally, for all intents and purposes, was an eternal and enormously powerful energy source.

But again there was a problem. It was clear from spectroscopic analysis that there wasn't enough radium, uranium, or any other radioactive substance in the Sun to turn it into a radioactive celestial torch. What radioactivity did do, though, was open the door to the possibility of obtaining energy from the atom. Not chemical energy from the bonds between atoms, but the incredible energy from an atom's nucleus. Rutherford figured out that radioactive substances were emitting an alpha particle—a helium atom stripped of its electrons—that was the atomic equivalent of a runaway freight train, with all its massive force and potential for damage. What would happen if alpha particles were fired at atoms? To find out, while the Great War raged across the English Channel, Rutherford fired alpha particles at nitrogen gas. The result changed the world and revealed a new kind of nuclear reaction: the alpha particles split the nitrogen atoms and, in doing so, transformed them. When the helium nucleus, with atomic weight 4, slammed into the nitrogen nucleus, with atomic weight 14, it fused with the nitrogen nucleus, spitting out a hydrogen atom (atomic weight 1) in the process. That hydrogen nuclei could be emitted as part of radioactive decay was amazing enough, but even more amazing was what happened to the nitrogen. It no longer had an atomic weight of 14 but rather one of 17. It was no longer nitrogen. It was oxygen. On realizing this, Rutherford is claimed to have bellowed at his lab assistant, "For God's sake don't use the word transmutation! They'll take us for alchemists." Whatever word he used, Rutherford had hinted at the deepest link between atoms and the stars.

The news of Rutherford's splitting the atom made headlines worldwide, but it didn't surprise his Cambridge colleague, the astronomer Arthur Eddington. For Eddington, it was the confirmation of something he already suspected: the Sun and the stars were powered by the transmutation of atoms. In person, Eddington was shy, and his Cambridge lectures were notoriously diffuse. "The problem," one student of his recalled, "was that Eddington had no proper connection between his brain and his mouth. As far as I could tell, he began in midsentence and stopped at the end of the hour, without any full stops between." For all his verbal maladroitness, given a pen and paper, Eddington had the voice of an angel. He was the greatest astronomy popularizer of the early twentieth century. His best-selling books, including *Stars and Atoms*, were wrapped and given as Christmas presents across the British Empire. Eddington, like many great physicists and astrophysicists, didn't just work through equations as equations. He had an intuitive, visceral sense of how mathematics applied to real-world objects—atoms, stars, and us.

When Eddington considered Kelvin's gravitational contraction model, he saw dim-witted thinking: "If the [gravitational] contraction theory were proposed today as a novel hypothesis," he told an audience in 1920, "I do not think it would stand the smallest chance of acceptance. . . . Only the inertia of tradition keeps the contraction hypothesis alive—or rather, not alive, but an unburied corpse." What was remarkable about stars, Eddington told the crowd, is not that they gravitationally contract, but that they don't contract *more*. Furthermore, astronomers had recently found stars that weren't getting smaller, but bigger. Recent observations of red giant stars showed that these stars were producing vast amounts of energy that puffed them out into distended versions of their earlier selves. These puffed-out stars didn't make any sense if a star's energy resulted only from gravitational collapse, in which case, a star's life was a one-way track of shrinking. "If we decide to inter the corpse," Eddington continued, drawing out his line of reasoning, "let us frankly recognize the position in which we are left. A star is drawing on some vast reservoir of energy unknown to us. This reservoir can scarcely be other than the sub-atomic energy."

In hindsight, with our homes powered by nuclear fission and our thoughts on the atom turned to avoiding the nuclear obliteration of human civilization, Eddington's claim seems pretty blasé. To most in his audience, however, it was wild speculation, and Eddington knew it. Yet, among astronomers at the time, he was geographically and intellectually closest to the seat of an emerging new world order. He told his audience about the work going on across the Cambridge commons at the Cavendish Laboratory. There, the physicist Francis William Aston had built one of the first mass spectrometers, the equivalent of a bathroom scale for atoms. For several years, and intensely since the end of the Great War, Aston had been extending Dalton's work of a century earlier by carefully weighing different elements. Aston's greater level of accuracy led to two new dramatic insights. First, not all atoms of an element are equal. Aston demonstrated the existence of hundreds of isotopes, atoms of an element with one or more additional neutrons. This in itself opened a crack in the idea of the perfect, immutable atom. But the accuracy of Aston's measurements added a kicker. Using Einstein's revolutionary insight into the equivalency of mass and energy—contained in the iconic formula $E=mc^2$, Aston's measurements enabled physicists to calculate each atom's equivalency in energy. For other scientists, the physical reality of this equation might have seemed more fantasy than fact, but Eddington was one of a handful of scientists in the world who'd seen the power of Einstein's relativistic predictions with his own eyes.

In May 1919, he'd led an expedition to the island of Principe, off the coast of West Africa, and during a solar eclipse had measured the gravitational bending of starlight by the Sun, the first observational confirmation of Einstein's general theory of relativity.

What was remarkable, Eddington told his audience, was that a helium atom, the second atom in the march of the periodic table, weighed just a little less than the cumulative mass of four hydrogen atoms. Aston had calculated the atomic weight of hydrogen as 1.0008, and that of helium as 4. If four hydrogen atoms were being fused together to form helium, what happened to the seemingly negligible 0.0032 atomic weight difference? It was raining down on them as light and heat from the Sun and stars, said Eddington. It might seem like a tiny amount of mass for such a huge job, but a small amount of mass is converted into an enormous amount of energy. Eddington calculated that if just 5 percent of a star is initially hydrogen, its gradual fusion into helium would give stars the power to shine for fifteen billion years. There was no doubt, he concluded, "that all the elements are constituted out of hydrogen." For the doubters in the audience, Eddington referred to his Cambridge colleague Rutherford and his transmutation of atoms: "What is possible in the Cavendish Laboratory may not be too difficult in the Sun."

Two decades passed before other scientists turned Eddington's wild speculation into conclusive mathematical description. By that time, atomic physicists had calculated hydrogen's and helium's precise quantum energy states and behaviors. With this in his metaphorical pocket, the German American physicist Hans Bethe—dubbed "the great problem solver" by his nuclear physics colleagues—solved a century-old riddle while waiting for dinner in a train car. His back-of-the-napkin calculations would lead to a 1939 paper published in the *Physical Review*, "Energy Production in Stars," the basis for his 1967 Nobel Prize. This was one of the last papers this refugee from Nazi anti-Semitism wrote before leading the theoretical division of the Manhattan Project, where he helped to build the world's first atomic bombs. In his paper, Bethe showed how stars get their energy: not by splitting atoms but by fusing them. A star is a balancing act between the gravitational forces exerted on matter, heating it so intensely that it fuses atoms, and the massive energy this process releases, pushing the ball of atoms outward. George Ellery Hale's dream had come true—atomic physics and the stars were as one. Bethe had convinced both physicists and astronomers that nuclear reactions fuel stars. His calculations showed that the Sun's luminosity was just as could be expected from the nuclear reactions at the estimated temperature at the Sun's core.

There was a critical wrinkle in Bethe's conclusion about stellar fusion. According to his calculations, stars quickly reached an elemental dead end. Bethe concluded that no elements heavier than helium could be formed in ordinary stars. The stars and the elements were joined, but only by the thinnest of threads—from hydrogen to helium. It appeared that stars could only account for the simplest of alchemical transformations, those from element 1 to 2. Bethe, "the great problem solver," had solved one line of the cosmic crossword, only to find that it intersected with long rows of other unknown letters. Now the hunt was really on for the origin of the elements.

BIG-BANG ATOMS

With the eruption of World War II, the world's leading physicists turned their knowledge of Nature's ways to aid the frenzied war effort, from fine-tuning the nascent science of radar to the elaborate mathematical cracking of secret codes and, in the United States, the atomic-bomb-building Manhattan Project. Yet, for many of them, the scintillating, unsolved mystery of the origin of the elements burned in the back of their minds. At the war's end, the question reemerged in a dramatic new light, now with three solid clues to guide the quest—clues that for the first time set firm, measurable limits on where to look, and a way to confirm or reject any potential answers. From these clues, two epic opposing views emerged in a clash that pitted the big bang against the stars.

The first clue was the question of temperature. Bethe had concluded that stars weren't hot enough to forge elements heavier than helium. He'd calculated that the temperatures needed to forge the elements heavier than helium required temperatures in the range of billions of degrees Fahrenheit, far hotter a temperature than was thought to occur in any star. When astronomers scoured the heavens, it appeared there was nowhere hot enough to cook up the elements—a cauldron hundreds of times hotter than the core of the Sun at twenty-eight million degrees Fahrenheit.

The second clue lay in a new, nuanced understanding of atoms and how they change, which had been provided by atomic weapons. The world's first nuclear bomb blast, on July 16, 1945, in the New Mexico desert, followed by the devastating nuclear attacks on Hiroshima and Nagasaki, Japan, were for astrophysicists terrible beacons pointing the way in the quest for the origin of the elements. As scientists shielded their eyes from a blast brighter than a thousand Suns, they were viscerally aware that they'd stepped into the new realm of the atom, one they could explore and test here on Earth. For the new breed of experimental astrophysicists, such as Caltech's William Fowler, the combination of Bethe's 1939 paper and his bomb unleashed a new view of what nuclear physicists were really doing. They now had their hands on cosmic powers in their labs.

The first computer models of stars were built on computer simulations of the atmospheric detonation of nuclear weapons—the closest events on Earth to what happens in stars—developed at the US Department of Energy's Lawrence Livermore National Laboratory. Experimental astrophysicists witnessed the transmutation of elements on a global scale: nuclear fallout. The same intense Cold War nuclear weapons testing that began in the mid-1940s and led to radioactive residues in breast milk around the globe eventually produced a slew of declassified data that, by the early 1950s, gave nuclear physicists the results of the elements forged in these atomic blasts. Nuclear scientists could also use the earliest particle accelerators, such as those at Caltech's Kellogg Radiation Lab, to test the burning question of how protons and neutrons got into and out of a nucleus and resulted in the transmutation of atoms.

While the temperature and details of nuclear cookery were important in unraveling the origin of the elements, the most important key to unlocking this secret lay in a third, jagged clue: a new graph not of types of elements but of their relative abundances. Judging by the equally apportioned Scrabble®-tile-like boxes that make up the periodic table, there is no hint that huge differences exist in the relative abundances of the elements. In the periodic table, each element gets the same-sized square, but not so in Nature, where there are enormous disparities. Until the early twentieth century, little scientific attention was paid to the relative abundances of the elements. The ancients knew that iron is much more abundant than gold, but not until the late 1930s—just as Hans Bethe was figuring out what makes stars shine—did the aptly named German geochemist Victor Moritz Goldschmidt reach for a cosmic perspective on the abundance of the elements—their *universal* abundances. To do this, Goldschmidt combined elemental data from terrestrial rocks and added data from extraterrestrial rocks—meteorites—which he believed might represent the average composition of cosmic matter. The resulting graph, the first detailed estimate of the cosmic abundance of the elements, struck all who viewed it as a cosmic riddle.

Unlike the even pattern of the periodic table, Goldschmidt's graph of cosmic elemental abundances is a saw-toothed jumble that gives the same impression of looking into the jagged interior of the Himalayas. Arranged by atomic number, from hydrogen on the left to uranium on the right, what's immediately clear is a steeply cascading saw-toothed pattern, in which hydrogen is an Everest, with a secondary helium peak, towering over a zigzagging flatland of the majority of the heaviest elements, from germanium to sparse uranium, far below. Between these two extremes lies a striking intermediate range of elemental peaks. After plummeting down to lithium, beryllium, and boron, the graph of abundances shoots skyward to carbon, climbs to peak at neighboring oxygen, before starting a zigzag descent, interrupted by a spiking peak at iron.

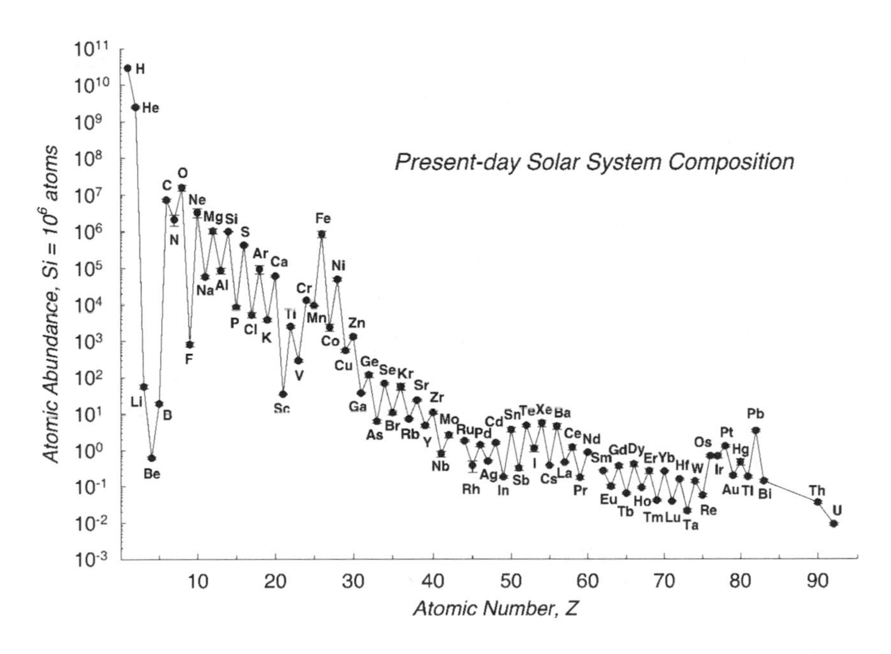

Figure 3.2. A modern graph of cosmic elemental abundances, as measured in our Solar System. *Image courtesy of Katharina Lodders.*

We now know that, taken together, all the elements that we think of as "stuff," from carbon to cobalt and potassium to plutonium, make up less than a measly 2 percent by mass of all the atoms in the universe—hydrogen accounts for about 75 percent, and helium for 23 percent. On an exam, it probably wouldn't make that much difference whether your mark is 84 or 86 percent. But in our cosmos, at least from our perspective, it's this 2 percent that makes all the difference. And among this 2 percent, there are vast differences in abundance. For example, in our Solar System, for every trillion atoms of hydrogen there are about 600 million atoms of oxygen and 33 million of iron; but there are only about 8 atoms of gold and 19 of silver, giving a cosmic context to bullion's enduring value.

Any theory of cosmic-element formation had to fit Goldschmidt's rudimentary map of the cosmic elemental abundances. There was a great cosmic story hidden in its atomic valleys and summits, but what was it?

In 1946, the war over, Russian émigré George Gamow thought he'd finally grasped the location of the Philosopher's Stone. To know the origin of the elements, researchers had to go back to another beginning: the origin of time. One of the first protocosmologists, the stocky and jocular Gamow, who'd landed a position at George Washington University in the American capital, reasoned that astronomers couldn't spot the celestial furnace that fit the bill for element forging because it no longer existed. The origin of the elements wasn't in the stars but in the fiery birth of the universe. The universe and the elements were born together.

George Gamow believed that all the evidence pointed to a massive primeval explosion—the then-controversial and unproven idea of the eruptive birth of the universe, or the big bang. Gamow's scientific upbringing had perfectly positioned him to make this conjecture. He'd earned his PhD in Leningrad under the influential mathematician Alexander Friedmann, who'd independently discovered that when Einstein's equations of general relativity were crunched, the results predicted that the universe had expanded from a much smaller beginning. Einstein initially balked at the idea, preferring the aesthetics of a static cosmos. But this prediction of general relativity was corroborated by Edwin Hubble's 1929 discovery that the galaxies are indeed racing away from one another; the ones farthest from us are moving at greater speed than those closest.

Gamow later forged his own reputation when he showed how the new theory of quantum mechanics could describe radioactive alpha decay—the phenomenon that occurs when an atom spontaneously spits out a helium nucleus. What was even more impressive was that he predicted that this quantum decay process also worked in reverse: that with enough energy, an alpha particle could burrow *into* a nucleus in a process called quantum tunneling, triggering artificial nuclear decay. Gamow had considered the nature of matter from the atom to the cosmos, and now he envisioned its emergence from a single creative event, meaning that all the phenomena we see and experience—time, space, and the varieties of matter—were created together in little more than the blink of an eye. It was an idea of enormous mythic resonance, a single act of creation that would later get an enthusiastic nod from Pope Pius XII as evidence of biblical veracity.

For Gamow, the origin of the elements was most important not in itself but for understanding the origin of the universe. The distribution of the elements was the remnant signal that indicated the universe's temperature, density, and rate of expansion at the beginning of time and space. What had it been like? The answer lay in the nature of the elements from hydrogen to gold and uranium. In September 1946, Gamow published an article titled "Expanding Universe and the Origin of Elements," which laid out what could be called his "Big Build-Up" hypothesis. Gamow argued that the key pieces of the origin-of-the-elements puzzle—temperature, nuclear transformations, and, above all, Goldschmidt's graph of cosmic elemental abundances—all made sense in the context of a super-hot (more than 180 billion degrees Fahrenheit), rapidly expanding moment of cosmic birth, during which the universe cooked up the contents of the periodic table in the first hour of creation.

The key to this, Gamow said, was a continuous building-up process arrested by the rapid expansion and cooling of the primordial matter. He painted a picture of the cosmos' first seconds as a scorching, dense soup of (free) neutrons glomming onto one another, building up larger and larger nuclei in not much more than a second, after which the soup was too cool and diffuse to squeeze together neutrons. After this period, there was a settling out as neutrons decayed into protons, releasing electrons and eventually forming the abundance distribution we see today. Therefore, there was little time for the heaviest elements to form—thus their low cosmic abundances as determined by Goldschmidt—and the distribution of the lighter elements was in part determined by a process of nuclear decay and settling.

The key nuclear physics was the gradual fattening-up of atoms from hydrogen to carbon and onward up the periodic table by the addition of progressively more neutrons. As neutral particles (those without a negative or positive charge), neutrons have a much easier time entering a positively charged nucleus full of protons than does another proton, which is forcefully repelled—as are the positively charged ends of two magnets—by what physicists call the Coulomb barrier. Gamow imagined the newborn universe awash in a tsunami of neutrons transmutating every atom in their path.

But as with so many contenders before Gamow, there was a problem, and this time it lay with atomic masses 5 and 8. And Gamow and his PhD student Ralph Alpher, who led much of the number crunching, knew it. A quick look at the periodic table reveals that there's no element with an atomic weight of 5 or 8. Atomic physicists also knew that no element had an isotope (a variant with more neutrons) that was stable for any length of time at this atomic weight. If the elements were built in a singular stepwise process up the ladder of atomic weights, a deadly gap existed at 5 and then again at 8. However, like many great scientists, Gamow didn't let these details destroy the central storyline of what looked like an elegant theory. Gamow and Alpher knew they had a core problem, but in the muddle-through approach that scientists often take, they chose to ignore it in public, in the hope that they'd resolve it in private in time to provide a convincing public explanation.

When the criticisms started, Gamow sidestepped them with his characteristic mirth, illustrating articles with images of himself energetically jumping ditches at atomic weights 5 and 8. But there was another problem. If the elements were created simultaneously, they ought to be evenly distributed throughout the cosmos. Yet astronomers were finding stars with widely varying amounts of heavier elements, up to a hundredfold difference from one star to another.

Gamow had time to look for an explanation. Most astrophysicists were taken with the notion that time and all matter had started as one in a single conflagration. But there was a man who thought otherwise. He didn't like the idea of this singular cosmic birth—dismissively calling it a "big bang"—and he hadn't given up on the stars as the birthplace of the elements. The stars didn't lack creativity; it was we who lacked it in our imagination of the stars. The man was Fred Hoyle, the scorned prophet of the Stardust Revolution.

LET THERE BE HOYLE

In his later life, it would be fair to say that a lot of fellow scientists thought that Cambridge astronomer Sir Fred Hoyle was a great mind gone far, far astray. Those less familiar with his enormous lifelong creativity and insight—or those just less forgiving of human foibles—simply called him a nut. Perusing Hoyle's prodigious and diverse body of work accomplished in the two decades before his death, it's easy to see why. The topics that filled his mind scream out like the garish cover of a supermarket tabloid. The astrophysicist with no fossil training took on the world of paleontology, claiming that the famous bird-dinosaur fossil Archaeopteryx was a Piltdown Man–like fake. He argued vociferously that disease outbreaks were caused by viruses from outer space. And his cosmological journey led him to believe that the only rational conclusion is that the universe was created through intelligent design.

In obituaries and biographies following Hoyle's death in 2001 at the age of eighty-six, writers seeking balance pointed out that Sir Fred Hoyle was one of the greatest astrophysicists and astronomy popularizers of the twentieth century. He captured the public's imagination in his evocative BBC radio broadcasts on astronomy and in bestselling popular-science and science-fiction books. He was the symbol of cosmic thinking for a generation, inspiring legions of budding astronomers. But Hoyle's professional career, from his very first professional presentation, was characterized first by controversy and later by outright conflict.

If you've never heard of Sir Fred, it's in large part because of what he was most famously wrong about: the origins of the universe. It was during one of his BBC broadcasts in 1949 that Britons gathered around their radios first heard of the "big bang"—a term the articulate Hoyle coined to describe the erroneous, he ventured, idea of a singular eruptive start to time and space in the distant past. Hoyle imagined, and for almost half a century championed, the dominant alternative twentieth-century story of the nature of the universe, one that envisioned an eternal cosmos. What's lost in this tale is that what Fred Hoyle was right about, and what he championed alone against great opposition, is just as central to the story of our cosmic origins. Our knowledge of the original cosmic birthplace of every atom in our bodies owes more to this straight-talking Yorkshireman than to any other person who ever lived. In fact, if there was a prophet of the Stardust Revolution, one who saw in the details a larger story of a living cosmos, it was Fred Hoyle. It is his story that carries us to the stellar origin of the elements.

From the time he entered Cambridge University as a math and physics student in 1933, Fred Hoyle—stocky, with curly black hair and a genial, forthright nature—dreamed of making a big find. He'd come from working-class origins; his father survived the trenches in World War I, but Hoyle defined himself not by where he came from but rather by his deep curiosity, tenacity, and intelligence. He wanted to discover something essential about the world, such as Newton's laws of motion or Einstein's general relativity. Upon his graduation in 1938, however, it looked like the door was slamming on Hoyle's dream before he'd even got started. He faced a problem of timing. He'd arrived at Cambridge at the tail end of the golden age of the quantum atom. A previous generation of atomic physicists, led by Niels Bohr in Copenhagen, had circumnavigated the weird and wonderful world of the quantum atom—its dual wave-particle nature, its precise energy states, and probabilistic nature.

On that very campus in 1932, Ernest Rutherford had discovered the neutron, rounding out the three-member ensemble of main atomic particles: a core of protons and neutrons surrounded by a jumping quantum field of electrons. Hoyle's thesis adviser, Paul Dirac, was part of this quantum coterie, adding key insights into the quantum nature of electrons. The positive side of this state of affairs is that Hoyle had been steeped in one of the greatest intellectual environments in Cambridge's four-hundred-year history. He'd learned quantum mechanics from Dirac and relativity from Eddington. But the bad news came from Dirac, when Hoyle sought direction about his future in science. Dirac told his student that the best berries had been picked in the field of quantum physics, and that the field would be fallow for decades to come. Another young scientist might have beetled away at secondary problems in quantum physics, content to eat the table scraps from the prior intellectual feast. Not Hoyle; if quantum physics was mined of big finds, then there were revelations to be uncovered elsewhere. He considered biology and astronomy and serendipitously landed on the latter. For in astronomy, young Hoyle's timing couldn't have been better.

During World War II, Hoyle was assigned to the intense effort to improve naval radar. But at the back of his mind were the stars. In 1944, at a conference on naval radar during his first trip to the United States, Hoyle went scientifically AWOL, taking the opportunity to visit America's leading astronomers and astrophysicists. At Princeton, he discussed the lives of stars with the preeminent American astronomer Henry Norris Russell, whose work had shown the fundamental relationship between a star's temperature and its energy output. Across the continent, at Mount Wilson, Hoyle spent hours intently listening to Walter Baade, a German refugee who'd used the wartime blackouts over Los Angeles to take the most detailed images yet of distant galaxies. Baade regaled Hoyle with his revolutionary observation of a massive star's explosive death—a supernova. Back in Cambridge at war's end, Hoyle took the stories from his travels and began to develop a radical new view of stars—the origins of not only heat and light but also of the elements. Tying together his knowledge of atomic physics and astrophysics, he began to develop a detailed view of stars as gradually evolving nuclear cauldrons, burning one nuclear fuel and then another, on a one-way path to their own deaths.

If there's a case study for great minds looking at the same information and coming to very different conclusions, it is the one concerning the cosmic origin of the elements. Fred Hoyle and George Gamow had both looked at Goldschmidt's data on the relative abundances of the elements, considered the same nuclear reactions and the particular temperatures and pressures involved, and even consulted the same scientific colleagues. While Gamow, in Washington, DC, was envisioning a singular birth for the elements, on the other side of the Atlantic, Hoyle was envisioning a radically different ongoing process of elemental creation. Within months of Gamow's 1946 paper, Hoyle published the first version of his view of the creation of the elements, a paper titled "The Synthesis of the Elements from Hydrogen." He acknowledged that transmuting hydrogen atoms into iron required temperatures hundreds of times hotter than normal stars. But, he said, there were places in the cosmos that were just right for cooking up the elements: massive exploding stars. If stars were static, their temperature remained static. But Hoyle took to heart what astronomers were observing—stars had lives. They died. And Hoyle calculated that as a dying star collapsed and exploded as a supernova, its temperature would skyrocket, sparking a flurry of nuclear reactions.

Critically, noted Hoyle, on the basis of his quantum mechanical calculations, this chain of supernova fusion reactions would reach a final peak at iron before plummeting to lower levels of element production, just as seen in Goldschmidt's table of elemental abundances. As elements grow larger, the repulsive force between their positively charged nuclei makes it increasingly difficult to squeeze them together. Iron, element 26, is the last element of the periodic table whose creation via fusing two smaller nuclei together releases energy. Fusing two atoms of iron requires, rather than releases, energy. Thus, what has come to be known as the "iron peak" is a critical juncture in the way stars make elements. The twenty-five elements before iron on the periodic table fuel stars, whereas those after iron rely on stellar energy to forge them.

For the first three months of 1953, Hoyle returned to Caltech and Mount Wilson as a visiting astrophysics lecturer, though he was still a voice in the wilderness when it came to his view of the origin of the elements. But he'd arrived in California chewing on an idea that soon solidified into a possible solution to the 5 and 8 gap, which had tripped up Gamow's big-bang-origin-of-elements scenario. Hoyle believed he knew how atoms in stars were maneuvering around this atomic gap to make the next element along the periodic table: carbon. And he knew the one man who could prove his conjecture: William Fowler. By this time, Fowler was the acknowledged master of experimental astrophysics, of mimicking in the lab the processes going on in stars. Fowler had made his mark early by helping Hans Bethe calculate the rate at which carbon and nitrogen in stars catalyzed the fusion of hydrogen into helium. Now Hoyle needed Fowler's help. When he arrived at Fowler's Caltech office, what made Fred Hoyle's request preposterous was that he'd shown up not just with the outline of an atomic process to cook up carbon in stars, but he also came with the exact energy required to do it, 7.65 mega electron volts. Speaking in his strong Yorkshire accent, Hoyle excitedly told Fowler and his Kellogg lab colleagues that the key to the carbon conundrum, and thus the doorway to all the elements, lay in the final, resonant notes of something called the triple-alpha process—the stellar atomic tango of three helium nuclei, or alpha particles.

Fowler was already familiar with the basic idea. He'd demonstrated that there wasn't a stable nucleus with a mass of 8. And a year earlier, he'd worked with Hans Bethe's student, Edwin Salpeter, to help show that in stars it was possible that under just the right conditions two alpha particles could fuse, forming beryllium, and then, in the blink of an atomic eye, before the unstable beryllium lost a neutron, these two alpha particles could gobble up a third helium nucleus to form carbon. But Hoyle showed that this triple-alpha reaction as calculated resulted in a minuscule amount of carbon production, certainly not enough to explain the abundance of carbon that astronomers saw in stars.

For Hoyle, this carbon conundrum wasn't in the stars and atoms but rather was a result of our view of them. The evidence was clear—the universe is chock-full of carbon, not to mention carbon is life's backbone element. In a variation on Descartes's "I think, therefore I am," Hoyle came to a simple yet radical conclusion: "I'm made of carbon, therefore stars must make carbon." The question was, how? This is where the energy value 7.65 mega electron volts was significant, Hoyle told Fowler. In order for stars to cook up lots of carbon, rather than have the nuclear reactions produce nitrogen or oxygen—the neighboring elements along the periodic table—there must be a specific carbon-formation energy zone—a carbon-forming quantum sweet spot. Hoyle calculated that this occurred if the triple-alpha dance reached a quantum-resonant energy of 7.65 mega electron volts above carbon's lowest energy level, or ground state, a level where, for a nanosecond, beryllium and helium nuclei formed a resonant energetic coupling, like two notes of an octave forming a resonant pair. From this quantum harmony, carbon was born in the hearts of stars. Critically, Hoyle predicted that this carbon sweet spot took place at too low a temperature for the reaction to tip toward making oxygen.

To Fowler, Hoyle's pinpoint prediction was wild speculation. At the time, predicting a quantum-resonant state involving three particles seemed akin to predicting the winning number in a national lottery. On a more personal level, Hoyle the astrophysicist was also telling the world's best nuclear experimentalists that they'd somehow missed identifying a crucial energy state in carbon. Though he initially rebuked Hoyle, Fowler eventually gathered his Kellogg lab colleagues, and, while behind his office door he privately told them that he thought the British astrophysicist was wrong, he let Hoyle explain his carbon hypothesis. Several of the Kellogg scientists, impressed by Hoyle's public lectures at Caltech and realizing that they were perfectly set up to experimentally test Hoyle's prediction, got to work doing just that. In less than two weeks, they'd turned Hoyle's hunch into transformational insight. Hoyle was right. There, in their accelerator, were carbon-12 nuclei spitting out alpha particles that carried with them the news that the carbon-12 atom that released them had an energy of exactly 7.65 mega electron volts.

Hoyle's success was a personal epiphany, one that eventually changed him as much as it did our view of the stars. He was the first to understand that stars make carbon, the elemental core of life, and that because of a finely tuned energetic nuclear dance, the doorway to all the heavier elements starts with this life atom. For Hoyle, it was no coincidence that carbon emerged in a narrow energetic window that, pushed just a smidgen further—to 7.19 mega electron volts, where oxygen would be preferentially produced—would produce a universe largely devoid of carbon and of life as we know it. Thus, Hoyle came to believe that far from terrestrial life being a cosmic fluke, the universe is finely tuned for life. It was a revelation that deeply shaped Hoyle's future views on a living universe and the idea of an intelligent cosmic designer.

More immediately, the discovery was a huge boost for Hoyle the astrophysicist. "It was the most rapid turnaround of people's views that I've ever experienced," he reflected, "my going from a voice in the wilderness to suddenly a very large number of people believing the idea." Hoyle returned to Cambridge with not just the stellar doorway opened to carbon but also with the tantalizing glimpse of the nuclear stairway beyond. While in California in 1952, he met Mount Wilson astronomer Paul Merrill, a quarter century his senior, and Hoyle teased him for making the startling discovery of technetium in stars. He knew Merrill wasn't looking to change the world. Hoyle, on the other hand, was in the midst of doing just that. Merrill's discovery was a dazzling confirmation that, for the upstart Hoyle, he was moving in the right direction.

Back in Cambridge, this Darwin of the origin of the elements set to work calculating and imagining the intersection of the nuclear-reaction pathway and the conditions in stars that could form the most common elements from carbon to iron. The result was his 1954 paper "The Synthesis of Elements from Carbon to Nickel," one of the landmark papers in the history of science. Here, in page after page detailing complicated series of nuclear reactions, Hoyle painted a new view of stars. Far from being static, they were gradually evolving nuclear cauldrons, burning one nuclear fuel and then another, bringing on their own deaths and seeding the universe with the elements of their creation. It was an intimate portrait of a large star's old age. Hoyle described a process in which large, aging stars, those we now know to be about eight times the bulk of the Sun and larger, gradually develop an onion-like structure of layers of different elements. For most of their lives, stars light up by burning hydrogen in their cores. The vast majority of stars we see, from our Sun to those twinkling in the night sky, are at this stage of life called the main sequence. But when much of the star's hydrogen core is fused into helium, the star gradually gravitationally contracts. This contraction boosts its core temperature, igniting the fusion of helium into carbon and oxygen. Thus begins a sequential series of midlife stages, or core evolution, in which the star burns through one nuclear fuel after another, its onion-like layers filled with the nuclear wastes from earlier stages.

Hoyle's vision extended the notion of stars as sources of energy to stars as *both* lighthouses and element factories. He established that these two roles were inextricably linked, that just as energy and matter are linked in $E=mc^2$, it's impossible to talk about energy in the cosmos without its relationship to the changing nature of matter at its deepest level, and that the twinkling of stars is light from the ongoing creation of the elements. But it would take one further great effort to finally tip the scales on the origin of the elements.

THE ASTRONOMER'S PERIODIC TABLE

Meeting in the basement of the Kellogg Radiation Lab that day in 1952, Fred Hoyle and William Fowler hadn't only figured out how carbon forms in stars. In that moment, they'd forged a strong mutual respect, a common purpose, and a deep sense of awe about their pursuit of the origin of the elements.

In the fall of 1954, Fowler moved his family to Cambridge on a year-long Guggenheim Fellowship and began the focused work of solving a cosmic puzzle for which the shapes of the pieces were resolving into clearer and clearer detail. His arrival marked another serendipitous connection. In Cambridge, he attended a lecture given by Geoffrey and Margaret Burbidge, a young astronomy couple who'd been measuring the abundances of elements in various stars and had come across some in which the heavier elements were far more abun-

▲ For University of Arizona astrochemist Lucy Ziurys, understanding the origin and evolution of life on Earth requires tracing our molecular heritage to the stars. *Photo by the author.*

Mount Wilson Observatory solar observer Steve Padilla's daily sunspot drawings are part of a vision that sees the study of the Sun as the first step in extending Darwinian evolutionary thinking to the stars. *Photo by the author.* ▼

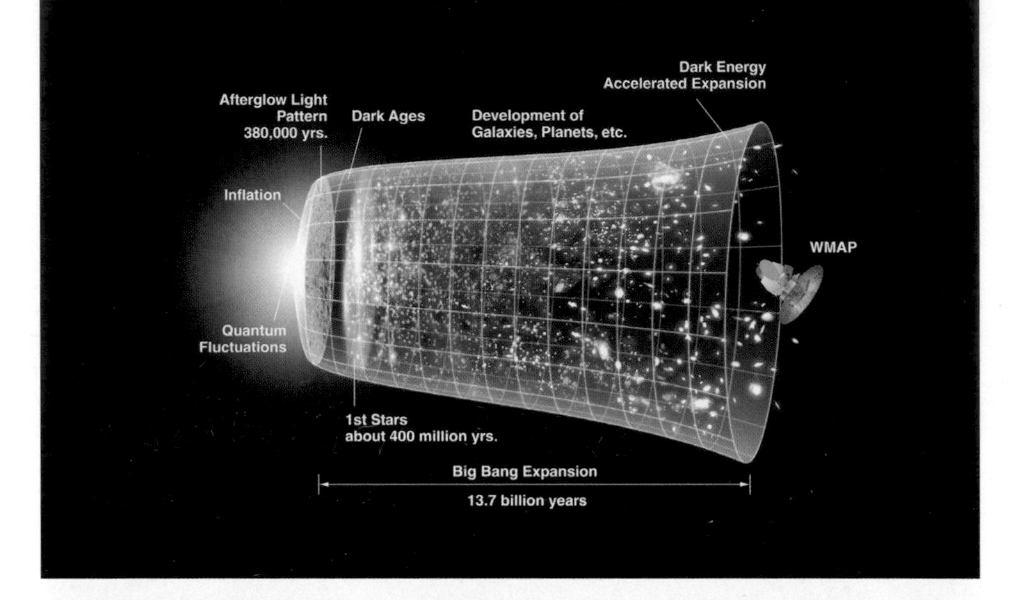

▲▲ Two views of cosmic history. The triumph of twentieth-century astrophysics was a vision of the expanding universe (*top image*). In the Stardust Revolution, this view is broadening to include a view of a biologically evolving cosmos, one in which we observe our stellar origins (*bottom image*). *Top image by NASA/WMAP Science Team. Bottom image by NASA Origins Program.*

▼▼ Stellar fingerprinting, or stellar spectroscopy, reveals what stars are made of. Joseph von Fraunhofer's spectrum of the Sun, hand-drawn in 1814, identified about three hundred of the thousands of Fraunhofer lines now observed—each line is one indication of an element in the Sun's atmosphere. *Top image courtesy of Fraunhofer. Bottom image courtesy of N. A. Sharp, NOAO/NSO/Kitt Peak FTS/AURA/NSF.*

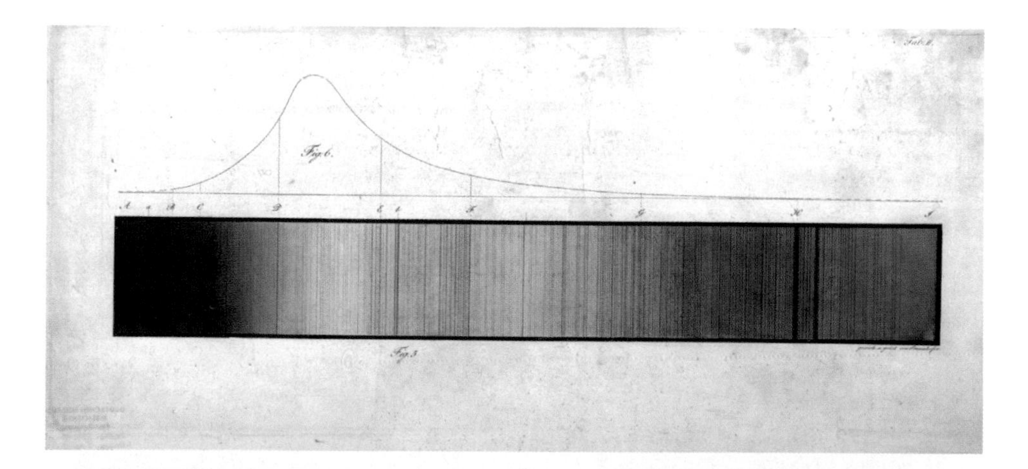

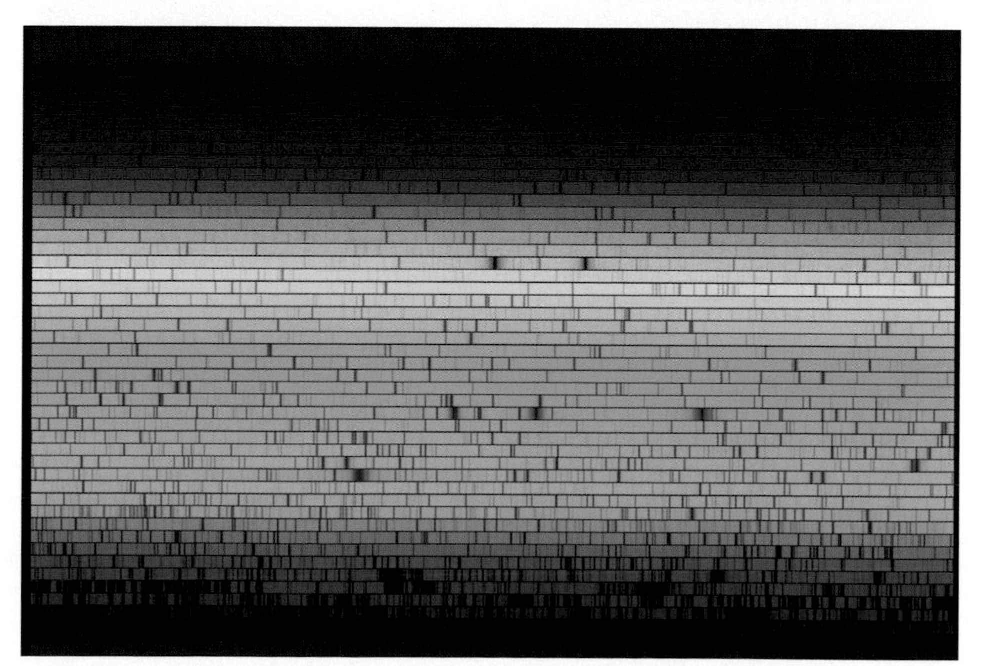

▲ Mount Wilson astronomer Paul W. Merrill's 1952 discovery of technetium in stars was the first observational evidence that stars are engines of element creation. *Photo reproduced with permission from the Huntington Library, San Marino, California.*

From left: Margaret Burbidge, Geoff Burbidge, William Fowler, and Fred Hoyle. The team that discovered that we are stardust, as explained in their 1957 paper "Synthesis of the Elements in Stars." *Photo reproduced with permission from the Master and Fellows of St. John's College, Cambridge.* ▼

▲ An educational version of the astronomer's periodic table, which connects the elements, and high-school students, to their stellar origins. *Image by NASA/GSFC.*

The 1936 book *The Origins of Life*, by Alexander Oparin (*seated*), inspired future generations of scientists to consider life's cosmic context, including Mexican origins-of-life researcher Antonio Lazcano (*standing*). *Photo by Antonio Lazcano.* ▼

▲▲ Two views of cosmic dust. The dusty lanes seen in edge-on views of distant galaxies, such as this one of the Sombrero galaxy (*top image*), prompted astronomers to see the dark lanes in our Milky Way—seen in 360-degree panorama (*bottom image*)—for what they are: cosmic dust. *Top image by NASA and The Hubble Heritage Team (STScI/AURA). Bottom image by ESO/H. H. Heyer.*

▲ Multiwavelength Milky Way. The discovery of infrared light opened our eyes to a multitude of new ways of seeing, each one of which provides a different view of our cosmic surroundings. *Image by NASA/GSFC.*

In Spitzer Space Telescope chief scientist Michael Werner's favorite image, Spitzer gets a clear view of dust's role in star birth and death in an area known as W5. *Photo by the author.* ▼

▲▼ The power of different ways of seeing. The Cigar galaxy, Messier-82 in visible light (*top image*), and as seen in the visible and infrared (*bottom image*), revealing vast clouds of dust and organic molecules spewing into intergalactic space. *Top image courtesy of N. A. Sharp/NOAO/AURA/NSF. Bottom image by NASA/JPL-Caltech/C. Engebracht (University of Arizona).*

dant than usual. Margaret was an acolyte of Fred Hoyle's, and both she and Geoffrey would be lifelong friends with Hoyle; Geoffrey always among his staunchest and most vocal defenders. Margaret was an astute observational astronomer who could tease a light fingerprint out of the faintest star and thus nail its elemental composition. Geoff, who resembled the actor Charles Laughton—famous for playing Quasimodo in the 1939 film version of *The Hunchback of Notre Dame*—was a theoretical physicist who enjoyed noting that he became an astronomer by marrying one. At the core of the team were Fred, the astrophysicist, and Willy, the nuclear physicist.

Figure 3.3. Joining atoms and stars. Fred Hoyle, the astrophysicist, looks heavenward in thought, while William Fowler, the nuclear physicist, calculates the netherworld of the atom. *Photo reproduced with permission from the Archives, California Institute of Technology.*

The four were gripped by the sense of being on a mission. They began by tying the quirk of elemental abundances known as Harkin's rule—that elements with even numbers of protons are slightly more abundant than their odd-numbered neighbors—to the stars. They realized that this zigzag pattern is the result of what they later described as the shuffling and reshuffling of protons and neutrons among atoms within a stellar cauldron. In this case, first an atom in a star captures a careening neutron, getting heavier yet not changing its fundamental chemical nature. Then, however, a neutron might beta-decay—a spontaneous process in which an unstable neutron spits out an electron and morphs into a positively charged proton. With one more proton, that atom is a whole new element. Here was a process like climbing the rungs of a ladder, in which the more stable, even-numbered proton atoms represented the rungs, and the less stable, odd-numbered proton elements represented the gaps.

In the fall of 1955, their research shifted to the other side of the Atlantic, to Pasadena, where the Burbidges visited, Geoff as a Carnegie Fellow. If they were going to provide conclusive proof of elemental buildup in stars, they needed solid stellar evidence. For that, they wanted to go up the mountain to use Mount Wilson's giant telescopes. The problem was Margaret, or, more precisely, that she was a woman. Mount Wilson's common living quarters were dubbed "the Monastery" because two things weren't allowed: whiskey (in fact all alcohol) and women. There was a solid glass ceiling when it came to the stars. Before Margaret Burbidge could help change humanity's view of the universe, she needed to crack another great barrier, and indeed she went on to become both the first woman Astronomer Royal in the post's three-hundred-year history and the first woman president of the American Astronomical Society. Fowler convinced Mount Wilson's director, Ira Bowen, to allow Margaret at the observatory, with conditions: the Burbidges had to use their own transportation; reside in segregated quarters in a small cottage heated by wood in a potbelly stove; and eat on their own, bringing their own food. Given all this, Margaret opted not to share a personal fact with any of the Mount Wilson staff—she was pregnant. She was worried that she'd be barred from some of the heavy work involved in moving the observer's ladder and hid the growing evidence under her heavy observer's coat until she was six months pregnant and had collected the data she wanted.

From Mount Wilson, the Burbidges descended to Caltech with what they'd sought. Combing through the astronomical literature, they'd chosen an old star named HD 46407, a so-called barium star, because of the abundance of the metal barium in its spectrum. The Burbidges carefully measured the abundance of any other heavy elements they believed the star was forming through gradual neutron capture and subsequent beta decay. Many of these are the rare-earth elements—including lanthanum, neodymium, and praseodymium—that today are the essential metallic hearts of our high-tech tools, from smartphones to the batteries in hybrid vehicles. The barium star's light fingerprint showed it was stuffed with these rare earths—up to ten times the amount measured in other stars. The Burbidges had seen with their own eyes what others were describing. For almost a decade, astronomers had begun to see that not all stars had the same percentage of elements heavier than helium, what astronomers refer to collectively as metals. Merrill's detection of technetium had been the smoking gun, and now the Burbidges had a roomful of stellar elemental forensics that all pointed in one direction: stars were changing themselves. They were maturing and, in the process, creating new elements.

The final piece of the cosmic puzzle came almost as a gift-wrapped Christmas present in the January 1956 issue of *Reviews of Modern Physics*, in an article titled "Abundances of the Elements." Here were the latest detailed calculations of cosmic elemental abundances and an updated version of Goldschmidt's chart—the equivalent of a puzzle-box cover against which the four-person stardust team could test their nuclear and stellar calculations to the decimal point. The authors, Hans Suess, a geochemist with the US Geological Survey, and Harold Urey, the father of planetary sciences and cosmochemistry at the University of Chicago, wanted them to do just that. They'd compiled an exhaustive list of elemental abundances from meteorite samples, stellar observations, and terrestrial samples to produce an abundance curve that is the template for today's estimates, all the way from hydrogen, down into the depths of lithium and boron, up through the iron peak, and down into the saw-toothed realm of the heaviest elements. Suess and Urey believed "the abundances of the elements and their isotopes reflected nuclear properties and that matter surrounding us bore signs of representing the ash of a cosmic nuclear fire in which it was created."

The Burbidges, Hoyle (who'd recently joined the others at Caltech), and Fowler set to work to prove just that. They worked to match the contours of the jagged line of the elemental abundances to the stellar forces that had shaped it. "Dear Hans," Fowler wrote on July 10, 1956, to Hans Bethe, his mentor and the man who'd solved the first riddle of what powers stars, "Hoyle, Burbidge and I are currently engaged in showing that the latest abundance curves of Urey and Suess are consistent with all schemes of nuclear cookery."

They titled their October 1957 paper in *Reviews of Modern Physics* simply "Synthesis of the Elements in Stars." In it they solved a millennia-old mystery, the origin of the elements from arsenic to zirconium and every letter in between. Astrophysicists, however, know it by another name, B^2FH—the acronym for the order of the authors, the two Burbidges, Fowler, and Hoyle. For astrophysicists, saying "B^2FH" is the equivalent of saying "Magna Carta" or "Rosetta Stone": the term conjures up much more than a single title offering incremental insight into the heart of Nature. It represents something monumental: a touchstone for a way of understanding the world, whether in relation to democracy, language, or our origins in the stars. B^2FH lives up to its billing just in length—at 104 pages, it's a behemoth that speaks more to a grand treatise than a single article, as if the Burbidges, Fowler, and Hoyle couldn't stop themselves in the telling the magnificent story they'd unraveled. Perhaps sensing the paper's import, Fowler, uncharacteristically for the author of a scientific paper, searched for quotes with which to preface it. He landed on another Willy, who wrote in *King Lear*: "It is the stars, The stars above us, govern our conditions."

B^2FH created a whole new language, that of the Stardust Revolution: a way of talking about the origin of the elements in stars, of stars as the cauldrons of atomic transmutation. The paper described six different key processes of stellar nucleosynthesis to account for the ninety naturally occurring terrestrial elements from hydrogen to uranium. First, there was hydrogen burning into helium, as described by Hans Bethe. Then, as a large star exhausted its hydrogen, it contracted and heated, and the helium would start its nuclear burn, forming carbon and oxygen. This set the stage for the alpha process, when massive stars' cores burn carbon and oxygen, forming the elements from neon to sulfur. Now the processes bifurcate, depending on a star's ultimate fate. In large stars that form red giants and gradually puff out their atmospheres, ending as carbon-oxygen cinders called white dwarf stars, there's a slow process of neutron capture, dubbed the s-process. The s-process gives new cosmic meaning to the word *slow*. In these elderly, extinguishing stars, neutrons stream out from the star's core, occasionally on just the right path and with just the right energy to collide with an existing atomic nucleus farther out in the star's shell and merge. Any given atomic nucleus doing a crazed energetic journey in a star captures a neutron about once every ten thousand years. In this geologically slow process of neutron capture and beta decay, atoms all the way up to element 83, bismuth, are forming as you read this.

Just as the s-process defines *slow*, the r-process, for *rapid*, gives new meaning to *fast*. Here, exploding stars—supernovas—provide the runaway energy and machine-gun spray of neutrons to accomplish in mere seconds what occurs in red giants over tens of millennia. It's as if there were two ways to bake a cake: one that was done just after you closed the oven door; the other that would be eaten by your distant descendants, who'd have created some mythic story to describe the bizarre cake that emerged, unbidden, from the oven. It's the r-process to which we owe the bulk of the world's precious metals, silver, platinum, and gold. From B^2FH we know that every gold wedding ring doesn't just bind a couple; through its trillions of gold atoms, each with seventy-nine protons and seventy-nine neutrons, it binds them to an alchemical cosmic blast that took place billions of years ago; that for weeks burned brighter than an entire galaxy of stars; and that, in the time it takes to say "I do," transmuted an Earth's mass worth of lesser elements into gleaming gold.

More than just elucidating the origin of the elements, B²FH created the astronomer's periodic table, one that would open previously unthinkable ways of seeing and understanding the cosmos. Like the director's version of a movie, the astronomer's periodic table provides the cosmic backstory. For astronomers, unlike chemists, their periodic table isn't a tool to understand how elements chemically combine. Instead it's a visualization of how these elements were formed. Thus every element is a tracer of its origins. Seen in this way, the periodic table becomes the story of the universe in ninety key symbols, from H for hydrogen to U for uranium. The elements form a kind of cosmic DNA. Each object—whether a speck of cosmic dust, a meteorite, a star, a galaxy, you, or indeed DNA itself—has in its elemental composition the detailed story of its cosmic origins.

Previous generations of astronomers used stars' light fingerprints to catalog stars, to group them, and to differentiate between them. However, they were stumped when it came to thinking in broader ecological terms of the changing relationships between stars over the billions of years of cosmic time. Until the understanding of stellar nucleosynthesis and the creation of the astronomer's periodic table, this was an intractable problem—there was nowhere to begin, no foothold of information from which to begin to reach for an understanding of cosmic natural history. The astronomer's periodic table changed this.

NOBEL CONCLUSIONS

B²FH has stood the test of time. In his 1983 Nobel Prize lecture, "The Quest for the Origin of the Elements," William Fowler noted that some of the elemental gaps in B²FH were filled in by others in the quarter century since; for example, how the small-element numbers 3, 4, and 5—lithium, beryllium, and boron—are created in interstellar space when larger atoms are shattered by cosmic radiation. Fowler also filled in another gap. He acknowledged his friend and colleague with precedence of discovery: "The grand concept of nucleosynthesis in stars was first definitively established by Fred Hoyle," he told the king of Sweden and others gathered for his lecture. Hoyle's former adversary, Gamow, had already long conceded defeat with a biblical-style ditty in which God forgets to create the elements in the big bang: "And so God said: 'Let there be Hoyle.' And there was Hoyle. And God saw Hoyle and told him to make the elements in any way he pleased."

By this time, Fowler had had time to reflect on the cosmic irony of it all. If it was indeed the stars that determine our fates, Hoyle's weren't twinkling but winking ironically. For one, Fowler shared the 1983 Nobel Prize in Physics with Subrahmanyan Chandrasekhar, one of the early supporters of big-bang nucleosynthesis, against which Hoyle had argued. This was a modest irony, compared with the bigger picture. While Hoyle continued to rail against the concept he'd named, his fastidious work in cosmic nucleosynthesis proved to be among the best, most concrete evidence for it. The existing amounts of hydrogen and, more importantly, helium provide tight constraints on the temperature, density, and rate of expansion in the birthing moments of the universe when these elements were cooked up. These primeval abundances were further refined by Hoyle, Fowler, and others with the 1965 discovery of the cosmic microwave background—the remnant birthing sounds of the big bang—which clinched the case of the big bang for most cosmologists.

When Fowler took the stage in Stockholm that December night, he'd already received Hoyle's response to his letter. "Dear Willy," Hoyle began in the little more than one-hundred-word reply to his colleague and friend of thirty years, "It has been clear I think to many people that the subject of nucleosynthesis in stars was likely to attract such an award sooner or later. In so far as I have thought about the matter myself, it had seemed to me that the two of us might share the Prize." The British papers thought the same, and headlines howled at the Nobel injustice. Journalists didn't have to look far for possible reasons the Nobel committee might have overlooked Hoyle. By this time, Sir Fred, always contentious, was well on his way to becoming a scientific pariah. From trumpeting the arrival from outer space of the virus that causes plagues, to the existence of an intelligent cosmic deity, Hoyle had isolated himself to the extent that the Nobel committee might well have shied away from giving him a global podium.

The experience only pushed Hoyle deeper in that direction. In his 1994 autobiography, *Home Is Where the Wind Blows: Chapters from a Cosmologist's Life*, Hoyle doesn't mention the Nobel Prize. His account of his scientific journey rings with the barely restrained subtheme of gross misattribution of credit in science, from Darwin getting credit for the earlier work of Alfred Russel Wallace and others, to Cambridge PhD student Jocelyn Bell's earlier snubbing by colleagues and the Nobel committee for her shared discovery of pulsars—rapidly spinning dead stars that emit intense streams of radio waves, blinking on and off like cosmic lighthouses. "The fiction is that credit is awarded in proportion to discovery," wrote Hoyle, "whereas, in fact, credit is awarded to those who influence the world, whether or not they were the real discoverers." At other moments, when it came to notions of success, Hoyle had a sense of philosophical resignation that ultimately success didn't really belong to anyone at all. It was all the voice of the cosmos. "Scientific discovery is not really what we think it is," he wrote, taking a different tack. "What happens is that the Universe programs our brains. The success really belongs to the Universe, not to us, and this has been the way of it throughout human civilization."

Indeed, Hoyle knew that he wasn't alone when it came to being overlooked in the attribution of credit for the discovery of stellar nucleosynthesis. "Don't forget Al Cameron," Geoffrey Burbidge said when I spoke with him two years before his death in 2010. Alastair Cameron was a Canadian nuclear physicist recently appointed as a professor at Iowa State University when he read and was entranced by Paul Merrill's famous 1952 "Technetium in Stars" paper. He beavered away in parallel to Hoyle, Fowler, and the Burbidges (contributing insight into neutron capture that was critical to Hoyle's 1954 paper), and almost single-handedly accomplished the equivalent of B^{2}FH, publishing two seminal papers in the same year, 1957. But since he was working at Canada's Chalk River Nuclear Laboratories, where a government wartime ethos persisted and extended secrecy to nuclear reactions in stars, his research was initially classified, and though it was later published, it remains in the deep shadow of B^{2}FH.

Whatever else happened, Hoyle, Fowler, the Burbidges, and a grand cast of other astronomers, physicists, chemists, and geologists across three centuries had put together the pieces to solve one of humanity's greatest puzzles—the origin of the elements. They'd traveled on an epic scientific journey that took them into the hearts of stars and way back to the origin of all that is. They'd tied together the periodic table of the elements, atomic physics, and stars into a new story of creation. In the end, they'd done more than any alchemist could dream. They didn't just uncover the Philosopher's Stone, the timeless wisdom of how to transmute elements from one to the other; they'd realized that we walk as the embodiment of this mystery. On Geoffrey Burbidge's death, his colleagues at the journal *Annual Reviews in Astronomy and Astrophysics*, which he'd edited for thirty-two years, offered this as part of a tribute: "When singer Joni Mitchell said 'We are stardust,' she was quoting Geoff."

For these scientists, after all this searching, the discovery of the origin of the elements wasn't so much an ending as a new beginning. They'd created the astronomer's periodic table, a way of exploring the family history of every atom and larger object in the cosmos. And B^{2}FH had tied the elements to the stars, just as others were working to understand another elemental relationship closer to home—that between the elements and life.

PART 2

THE INVISIBLE UNIVERSE

CHAPTER 4

THE ATOMS OF LIFE

It is a slightly arresting notion that if you were to pick yourself apart with tweezers, one atom at a time, you would produce a mound of fine atomic dust, none of which had ever been alive but all of which had once been you.

—Bill Bryson, *A Short History of Nearly Everything*, 2003

DARWIN'S GAP

You'd be hard-pressed to find a group of people with more experience thinking about the nature of life than the dozen or so scientists sitting around the table for the fall 2010 meeting of the Committee on the Origins and Evolution of Life of the US National Academies' Space Studies Board. The group, assembled at the former summer home of a Wall Street banker on the outskirts of Woods Hole, Massachusetts, runs the gamut of life research. There's a paleontologist who studies the billion-year-old fossils of the Earth's earliest multicellular organisms. There are microbiologists who show slides of wonderfully exotic locales where they've searched for extremophiles—single-celled creatures that make a living where, only decades ago, we thought life couldn't possibly get a foothold: in (not on, or under, but *in*) Antarctic ice, in sulfuric hot springs, or two miles underground, where, without light or oxygen, bacteria breathe sulfur and eat radioactivity. The meeting includes guests from the Craig Venter Institute, who describe their efforts to create artificial life. Venter made his fame and fortune inventing the technology that underpins all genomics research, including the Human Genome Project. His eponymous institute, which he heads, is leading the effort to make life from scratch. Rounding out the group are NASA scientists involved in looking for life on Mars and Saturn's moons.

Although its main task is to think about how and where life happens, this session begins with a presentation about how difficult it is to get rid of some undesirable forms of life. Via teleconference, a NASA official explains the challenges of sterilizing robotic spacecraft before their launch to distant locales in the Solar System. NASA's first life-science challenge in missions to the Moon and Mars isn't finding alien life but avoiding the transfer of microscopic stowaways from Earth and thus leaving our own life footprint elsewhere in the Solar System. As a partial fix for this problem, NASA uses ultraviolet (UV) light in its prelaunch clean rooms to sterilize spacecraft as part of its Planetary Protection Program. In what is perhaps the first case of NASA-driven bacterial evolution, this process has selected for UV-tolerant bacteria. In an effort to kill them, the space agency made them better and stronger. "This is the Earth. We can't keep the bacteria out," the NASA official complains.

The day's first lesson: life is tenacious. For most people today, the daunting problem with life isn't to know how it began but how to end it—from homeowners plotting ways to kill dandelions, to NASA engineers trying to rid *Deinococcus* bacteria from spacecraft—which is what creates the slight cognitive dissonance in the meeting room. This is, after all, the Committee on the *Origins* and Evolution of Life. Yet, often, the elephant in the room during the discussions is the question not of life and its evolution but of life's very beginnings.

It's a split that began with the great theorist of evolution, Charles Darwin. Some of Darwin's first readers, though awed by the observations in his 1859 masterwork *On the Origin of Species* and agreeing with his sweeping conclusions, couldn't help but notice a gap. While he artfully laid out the evidence for how life changes over time, Darwin didn't address the essential question: How did life start? One of his biggest fans, the eminent German natural philosopher Ernst

Haeckel, commented on this issue in 1862: "The chief defect of the Darwinian theory is that it throws no light on the origins of the primitive organism—probably a simple cell—from which all the others have descended."

Darwin's gap wasn't an oversight; it was a choice. Darwin knew he was in for a formidable scientific battle with the theory of evolution by natural selection, a topic on which there was voluminous living and fossil evidence from around the globe. The origin of life, however, was a completely different story. Darwin was fascinated by the question but believed at the time that the pursuit of the evolutionary origins of life in the distant past was beyond the scope of scientific testing. As such, any comment on the topic was sheer speculation, which would only invite equally insubstantial rebuttals, muddying the waters of his evolutionary argument. In a letter to his close friend Joseph Dalton Hooker in 1871, he put it more bluntly: "It is mere rubbish thinking, at present, of (the) origin of life; one might as well think of the origin of matter."

A century and a half after Darwin, many of the members of the Committee on the Origins and Evolution of Life still feel the same way. A number of evolutionary biologists strongly believe that it's impossible to unravel the series of events that led to life on Earth. In essence, they believe, the evidence is lost in the mists of time. Unlike dinosaur fossils, the four-billion-year-old organic evidence from the dawn of life has been destroyed, degraded, or changed so much that its earlier form is unrecognizable. Life has erased its own trail. Alternatively, some believe the emergence of terrestrial life was a unique, one-off, random event that's much too complex to experimentally reproduce. There was, however, one scientist in the meeting room whose view of the origins of life on Earth was shaped by a lesser-known great book of science, one that argues that tracing our family tree back to origins requires looking to the stars.

Antonio Lazcano is an outlier on the Committee on the Origins and Evolution of Life in more ways than one. A Mexican and a biology professor at the Universidad Nacional Autónoma de México, he's one of the few non-Americans ever to sit on the committee. Of medium build and with salt-and-pepper hair, Lazcano has a languid intensity, seeming to be perennially engaged in whatever the topic of conversation is. As a lunch mate, he's as happy discussing—and at times making lengthy pronouncements on—the arts, books (he's encyclopedic about Mexican culture), politics, or the history of science as he is his tablemate's latest research.

After several days of listening to presentations, Lazcano is still eager to talk. He sleeps only five hours a night and, when not sleeping, is a peripatetic thinker and autodidact on his favorite topic—how life on Earth began. He's the former president of the foremost group on the topic, the International Society for the Study of the Origin of Life. For Lazcano, our origins matter deeply when it comes to understanding ourselves. Evolution is, at its core, about the historical contingency of life, and to Lazcano, history means genealogy. Who, or what, begat whom? In this context, Darwinian evolutionary theory ties us to the Earth. It's the origin of life that ties us to the cosmos.

In his quest to understand our extreme genealogy, Lazcano is a great connector, someone whose life as well as his research brings the past closer. When I sit down with him for a drink in the aptly named Fuzion Restaurant, I realize that Charles Darwin is not that far away. Through Lazcano, I'm only four degrees of separation from the giant of evolution himself. And it all began with a book. "As a child, I'd always wanted to be either an astronomer or a chemist," says Lazcano, between sips of mint tea. "Then, when I was about eleven my father gave me a copy of Oparin's *Origin of Life*." It was a book that changed both Lazcano's life and the course of the Stardust Revolution.

ON THE ORIGIN OF LIFE

A century before and an ocean away, Alexander Ivanovich Oparin sat inside a Moscow apartment's sitting room, listening with rapt attention to the impassioned interjections of young university students like himself crammed into the too-small space. He listened most closely when the older man, Kliment A. Timiryazev, spoke. It was 1916, and there had been violent demonstrations on the cobbled streets outside, and the whispers that the czar's days were numbered had grown into shouts of defiance. But those gathered in Timiryazev's salon weren't plotting political revolution; they were galvanized by another great revolutionary way of thinking—evolution.

Timiryazev was Darwin's Russian bulldog. Since the publication of *On the Origin of Species*, Timiryazev, a leading plant physiologist at the University of Moscow, had grown to become the leading Russian proponent of a revolutionary new way of thinking about life's unfolding. As a young scientist, he had made an impromptu pilgrimage to the elderly Darwin's home, surprising his intellectual hero but nonetheless being invited to tea. Timiryazev was also an early Marxist. As such, his salon sessions married the ideas of inevitable biological change with those of irrepressible social change. Both evolution and dialectical materialism—a material rather than divine process of change based on competition and contradictions within a particular social environment—looked to history and competition as the basis for understanding biology and society. They viewed both as moving toward a better world.

When Oparin heard Timiryazev for the first time at a public lecture in Moscow, he was galvanized by the evolutionary firebrand. Already interested in botany and agronomy—his father worked in textiles and dyes—Oparin enrolled at the University of Moscow. Timiryazev had resigned his post at the university in 1911 in protest over the Ministry of Education's violent suppression of a student protest. In the great tradition of the day, Timiryazev continued to teach in the salon of his Moscow home, where Oparin and others gathered to listen to the great man.

At the university, Oparin studied the burgeoning field of biochemistry at a time when biologists and chemists were crossing one another's boundaries in pursuit of an understanding of life as a molecular, chemical phenomenon. They debated the molecular nature of the genetic material and whether bacteria even contained genetic material—most believed they did not (they do). As a student in the laboratory of renowned biochemist Alexei N. Bakh at the university's Karpov Physicochemical Institute, Oparin was at the heart of one of the world's foremost research labs, where researchers were teasing apart and identifying as best they could the molecular components whose interaction we call life.

In 1917, as the Russian Revolution reached its bloody climax, a twenty-three-year-old Oparin, steeped in the latest biochemistry and evolutionary theory, stepped out into an amazing and turbulent post-czarist society that was pulsating with competing visions for a new society. While Hubble was at work on Mount Wilson observing distant galaxies—work that would provide the underpinnings of the big bang—Oparin was contemplating biological beginnings. The young Russian biochemist was drawn to the big question of life's essential nature at a time when age-old answers—just as age-old forms of social rule in the form of czars and counts—no longer fit the times. A young man full of revolutionary zeal and new scientific insights, Oparin was ready to go where Darwin had feared to tread. Following the revolution, the streets and salons of Moscow were filled with pamphlets and booklets extolling new social and political visions. In November 1923, one of those revolutionary tracts, by A. I. Oparin, was ambitiously titled *The Origin of Life*.

THE SPONTANEOUS-GENERATION DEBATE

At first, Oparin's manifesto was lost in a tumultuous sea of competition. By 1936, however, his treatise had matured into a Russian masterwork, *The Origin of Life*, creating a buzz among biologists who could read Russian. Meanwhile, Oparin, dressed daily in a bow tie and with a crisply trimmed, Leninesque beard, had become the associate director of the Biochemical Institute of the USSR. Two years later, a fellow Russian working at a US university translated Oparin's treatise into English as *Origin of Life*. It was immediately recognized in the United States as a definitive tome, one that a *New York Times* book reviewer concluded was a "landmark for discussion for a long time to come." The reviewer was right. Oparin's book has defined the modern discussion of the origin of life because it wasn't merely filling a gap left by Darwin. It was also addressing one of the greatest, and largely forgotten, scientific debates of all time: the spontaneous generation of life.

The fact is, as Oparin notes in the first pages of his treatise, if you'd talked to leading natural philosophers throughout most of history, from Aristotle to Newton and on to Lamarck, the origin of life wasn't a troubling problem of the distant past; it was a present-day no-brainer. Life happened spontaneously all the time. You only had to look at the abundant daily anecdotal evidence—life sprouting unbidden at every turn from places where none existed before, whether it was mold on days-old bread or mushrooms on the forest floor. "Even in our own time of crowning achievement in the natural sciences," Oparin wrote, "the layman of civilized European countries not infrequently believes that worms are generated from manure, that the enemies abounding in the garden or field, the various parasites operative in our daily existence, arise spontaneously from refuse and every sort of filth." It's remarkable to think that ours is the first century (if not, then only the second) in human history in which the majority of people do not believe in the spontaneous generation of life.

However, not all protobiologists adhered to a belief in life's daily re-creation. For more than four hundred years, an intermittent back-and-forth debate simmered. In an age that had no awareness of microbes, the experimental evidence alternately appeared to support one side and then the other, with spontaneous generation usually carrying the day. Savants, theologians, and alchemists attributed various causes to the spontaneous generation of life, at times divine; at others, simply part of the eternal, natural order of the cosmos. But in all these cases, life was always a distinct, mysterious, secret force apart from the rest of matter. It was the *force vital.* It was a centuries-long debate that Darwin wanted to avoid.

The vision of an essential, unbridgeable gulf between nonlife and life was formalized into modern chemistry in the first half of the nineteenth century by the Swedish chemist Jöns Jacob Berzelius. Berzelius, who developed the system of element symbols used in the periodic table, saw a firm line between the chemistry of life and nonlife, and as a result he saw clear limits of what chemists could concoct in the laboratory. "Art cannot combine the elements of inorganic matter in the manner of living nature," he asserted in 1827. According to Berzelius, it was impossible for chemists to create "living" substances such as sugars, fats, and proteins. He coined the term "organic chemistry," meaning life chemistry, as distinct from inorganic chemistry, thus forging a dualist language that still haunts the teaching of chemistry and biology today.

Even as Berzelius was writing about the unbridgeable gap between the molecules of life and all others, a friend was proving him wrong. The first bridge between the molecules of life and nonlife came from synthesizing the main ingredient of urine in a test tube. Urea is the main waste product in urine. In fact, it is the reason we pee. Urea is our body's way of getting rid of otherwise toxic nitrogen leftovers from metabolism. It was clear to nineteenth-century chemists that urea was a product, albeit a waste product, of life. So, when in 1828 the German chemist Friedrich Wöhler created urea in laboratory glassware, other scientists were impressed. Wöhler heated ammonium cyanate to produce copious amounts of synthetic urea, chemically identical to the stuff in urine yet without the need for beer, coffee, or kidneys. This event prompted a profound realization that chemists are still absorbing: there's no chemical difference between, for example, the biologically produced methane in a cow's fart and the abiotically formed methane in the atmosphere of Jupiter. Both methanes are a single atom of carbon combined with four atoms of hydrogen.

Within a year of Berzelius's drawing a line in the sand between the realms of the inorganic molecules and life molecules, Wöhler showed—and other chemists after him also soon demonstrated, with a chemist's cupboard of molecules from sugars and amino acids—that biological molecules could be created from nonliving material. This directly challenged the then-predominant understanding of spontaneous generation that life arose from the self-organizing of previously living matter that alone carried vital force. Wöhler's results showed that life molecules could be created from what was thought of as matter that lacked the *force vital*. His achievement, however, was hardly a clean flush to the notion of the spontaneous generation of life. Instead, it helped set the stage for a showdown between the warring scientific camps.

In Paris in 1859, the same year that *On the Origin of Species* was published, one of France's most prominent natural philosophers tried to end the debate with a decisive blow. Félix Archimède Pouchet, director of the Natural History Museum in Rouen, published a massive, seven-hundred-page treatise purporting to experimentally prove spontaneous generation. Pouchet described his experiments, like those done by earlier scientists, which demonstrated that heated broths of organic matter, left to sit for several days, spontaneously became cloudy with microscopic life. Evidently, Pouchet argued, under divine direction, a vital force was at work, and organic particles carrying this life force were self-organizing into microbes.

Pouchet had thrown down the gauntlet to any opponents, and the next year, the French Academy of Sciences officially refereed the intellectual duel. It offered a 2,500-franc prize to whoever could shed new light on the question of spontaneous generation. The winner, announced in 1862, was none other than Louis Pasteur. Today, we think of the legendary French chemist and bacteriologist every time we drink a glass of pasteurized milk—milk that's been heated and then rapidly cooled to kill any harmful bacteria. But each glass of milk could equally give us pause to think about the origins of life, for Pasteur's sterilization work was ultimately driven by the debate over the spontaneous generation of life. From his ample experience with microorganisms through work in the French beer and wine industry,

Pasteur had a hunch that what Pouchet described as divine intervention is what we today call dirty lab ware and exposure to airborne contamination. At the time, there was no general understanding of the amazing ubiquity of microorganisms—bugs, in the vernacular; bacteria and viruses, to be more specific. While telescopes were revealing a new view of the heavens, microscopes were about to open our eyes to the invisible realm of the microbes, in the air and on surfaces, that cause disease.

Pasteur needed to prove that Pouchet had somehow inadvertently seeded his samples with invisible life and that an organic broth—such as an infusion of hay, when properly sterilized with heat—would remain lifeless. Pasteur had flasks made with various S-shaped necks that would allow the flow of air into the flasks, but not the entry of microbe-carrying dust, which would settle out and be trapped in the lower curve of the S-neck. In earlier work, he'd filtered and captured airborne dust particles, demonstrating that the dust carried the seeds of life. Now he showed that a broth added to one of these flasks and subsequently sterilized by heat remained clear and lifeless for months. However, in flasks in which the S-neck was broken, microbes quickly populated the broth, indicating that the microbes had entered from outside the broth.

In 1861, Pasteur published his results in an essay, "*Mémoire sur les corpuscules organisés qui existent dans l'atmosphère*" (loosely translated as "A Memoir on Organized Corpuscles in the Air"), which not only gave a fatal blow to the theory of spontaneous generation of life but laid the cornerstone for the modern germ theory: that diseases don't emerge spontaneously but are spread via the transfer of microorganisms. In his triumphal lecture at the Sorbonne in 1864, Pasteur proclaimed that spontaneous generation was dead: "Never will the doctrine of spontaneous generation recover from the mortal blow of this simple experiment." The next year, well on his way to hero status in France, Pasteur pocketed the 2,500-franc prize—though legend has it that, given his developing understanding of germ exposure, Pasteur refused to shake hands when he received the prize.

AN ELEMENTAL VIEW OF LIFE

For seventy years after Pasteur's definitive discovery, the question of the origin of life on Earth struggled to find a solid grounding. Some scientists, such as the famous Lord Kelvin—a lifelong opponent of evolutionary theory—found comfort in the idea that life was simply eternal. Others, such as the Swedish chemist Svante Arrhenius, proposed that the seeds of life on Earth arrived by cosmic delivery from elsewhere in the universe—the concept of *panspermia*, or "life everywhere," which avoided the question of this universal life's origins.

Into this gap came Alexander Oparin and his gem of a book *Origin of Life*. For Oparin, panspermia and eternal life were back-steps into a nineteenth-century "pit of vitalist conceptions," both lacking a physical, scientific framework. Oparin sought to understand life not as something that occurs in spite of the laws of physics and chemistry but as something that occurs as a result of them. "The problem of the origin of life cannot be solved in isolation from a study of the whole course of the development of matter which preceded this origin," Oparin wrote. "Life is not separated from the inorganic world by an impassable gulf—it arose as a new quality during the process of the development of that world."

In this sense, Oparin was continuing a line of dawning awareness of life's elemental nature begun a half century earlier. In 1869, when, in retrospect, one might be surprised that even the chambermaids and coachmen weren't debating the origin of life, Darwin's British bulldog, Thomas H. Huxley, president of the Geological Society of London, professor of natural history in the Royal College of Surgeons, and renowned orator, published an influential article that made just the above point. Titled "On the Physical Basis of Life," it argued that the nature of every living being is in fact the expression of what could be called the four great elements of life—carbon, nitrogen, oxygen, and hydrogen. What we call "life" isn't some mysterious *force vital*, Huxley said, but rather is the synergistic result of a myriad of combinations of these and other less abundant elements. Huxley made the point that when oxygen and hydrogen combine to form water, "we do not hesitate to believe that, in some way or another, [water's properties] result from the properties of the component element of water. We do not assume that something called 'aquosity' entered into it."

Though the popular notion is often that life is something fundamentally apart from the rest of the universe, when taken to its atomic essence, we, and all other life, from *E. coli* to elephants, are not a random, ad hoc, or odd assemblage of atoms but rather standard issue. It's a realization that triggered intellectual alarms for the late nineteenth-century astronomer William Huggins, the first person to document stars' elemental compositions. When he looked at his spectra, identifying what the stars are made of, he saw not foreign worlds but his own. "It is remarkable that the elements most widely diffused through the host of stars are some of the most closely connected with the constitution of the living organisms of our globe, including hydrogen, sodium, magnesium, and iron," he reflected. Parsed to our atomic essence, we look strikingly similar to the astronomer's periodic table, dominated by the same handful of elements that comprise the universe. Indeed, the four most common atoms that make up you and me—oxygen, carbon, nitrogen, and hydrogen—are also four of the five most common elements in the cosmos. (The fifth is helium, the second-most common element in the cosmos. It probably wasn't used as a life building block because it's an elemental loner—it doesn't readily bond with other atoms.) Take yourself apart atom by atom, so to speak, and you'd find that by mass four simple atoms—elements 1, 6, 7, and 8—compose, remarkably, approximately 96.2 percent of you. You'd end up with a pile of oxygen atoms that weighed in at about 65 percent of your total weight, followed by 18.5 percent carbon, 9.5 percent hydrogen, and, finally, 3.2 percent nitrogen. Of course, you're wondering about the other 3.8 percent of you, of which about 1 percent is phosphorous and sulfur, and 2 percent is salts such as sodium, calcium, and potassium. All the other twenty or so elements that make us who we are, from iron to copper and every other essential element in our morning vitamin, make up less than 0.03 percent of our body mass.

MOLECULAR EVOLUTION

Alexander Oparin took the reflection on life's elemental nature one critical step further. He turned to astronomy to understand biology. In the midst of his day job—isolating and characterizing enzymes at the University of Moscow—Oparin contemplated the origin of these complex molecules as the result of a process of planetary evolution. As a result, his office desk was strewn with papers offering the latest results not only in biochemistry but also in geology and most notably in astronomy. He pulled various concepts from these disciplines together to create a cornerstone of the Stardust Revolution and, arguably, became the first astrobiologist.

Oparin's great contribution to the Stardust Revolution was the realization that the magnificent story of the origin of life on Earth only becomes accessible when viewed in a cosmic context. In this perspective, life isn't magic; it's elemental, chemical, and cosmic. Oparin situated life in a universal, physical context in the same way that Newton's description of the universal laws of gravity joined heaven and Earth. He saw life as an evolutionary process embedded in cosmic evolution.

"The carbon atom in the Sun's atmosphere does not represent organic matter," Oparin wrote, "but the exceptional capacity of this element to form long atomic chains and to unite with other elements, such as hydrogen, oxygen, and nitrogen, is the hidden spring which under proper conditions of existence has furnished the impetus for the formation of organic compounds."

At this point you might think, "Isn't this just another version of the spontaneous generation of life, and hasn't that idea been trashed?" It's a good point, and one that Oparin was keen to clarify. The difference in his proposal is that he was not suggesting the holus-bolus appearance of single-celled organisms such as bacteria. Prior to the development of biochemistry (and even then), *small* was conflated with *simple*. Thus, it appeared plausible that a single-celled beast could self-organize from, for example, a sterilized broth of chicken soup. Oparin, a lab biochemist, was acutely aware of the chemical complexity of even the smallest life. "It must be understood," he pointed out, "that no matter how minute an organism may be or how elementary it may appear at first glance, it is nevertheless infinitely more complex than any simple solution of organic substances." On the same grounds, Oparin rejected the long-dominant idea that the first life was green. He knew that the chemical pathways and molecules involved in photosynthesis are far too complex to have been life's starting point.

Oparin set the framework for modern thinking about the origin of life on Earth by arguing that life didn't emerge spontaneously but rather that it came about through a long, gradual process of molecular evolution. Whereas Darwin thought about the gradual transformation of species of animals, Oparin, the biochemist, pondered the gradual evolution of species of molecules. "All these difficulties [related to the appearance of cellular complexity] disappear," Oparin argued, if we "take the standpoint that the simplest living organisms originated gradually by a long evolutionary process of organic substance and that they represent merely definite mileposts along the general historic road of evolution of matter." In this context, Oparin argued, the key question is: What was the Earth like in the distant past, when life first evolved from nonliving matter? The short answer is: very different. That, said Oparin, makes all the difference. Life on Earth didn't evolve on the planet as we now know it but on a primitive orb that we wouldn't recognize as home.

First of all, there was no life. On the face of it, this might seem facile, but it turns out to be crucial. Ironically, Oparin conjectured, the first protean life—whatever it might have been—could get a foothold precisely because there was nothing else around to eat or chemically degrade it. "However strange this may seem at first sight, a sterile, lifeless period in the existence of our planet was a necessary condition for the primary origin of life."

It was this line of thinking about the Earth's cosmic origins, about the Earth as a planet with its own history that marked the watershed in Oparin's stardust science. He analyzed hot-off-the-presses results from stellar spectroscopy, meteorite analysis, and the study of the atmospheres of other Solar System planets to come to a paradigm-shifting conclusion: Earth's atmosphere, and thus life, started without molecular oxygen. Today we think of oxygen as essential to life, and with good reason: spend more than a minute or two without oxygen, and we'd be worried not about the origin of life but rather about our imminent demise. Oxygen is the breath of life, both for humanity and for much of current life on Earth. But much of life on Earth is also anaerobic—from the bacteria in your gut to fermenting yeast. For these creatures, oxygen—a voracious electron stealer, or oxidizer—is deadly. We now know that oxygen in the atmosphere didn't fuel life but rather that it is in fact a result of life.

In planetary atmospheric chemistry there are two dominant types of atmospheres: those in which oxygen abounds and readily oxidizes all other atoms; and those dominated by hydrogen, which, rather than taking electrons prefers to share its sole electron. Oparin argued that rather than being oxidative, the early Earth's atmosphere was reducing, or rich, in hydrogen. This protean atmosphere was dominated by molecules bound with hydrogen. The evidence lay in the stars, meteorites, and other Solar System planets. Oparin noted that stellar spectroscopy revealed simple carbon-carbon, carbon-nitrogen, and carbon-hydrogen molecules in the atmospheres of stars. Similarly, the minerals in meteorites revealed that many were formed in environments without oxygen. Finally, the recent spectroscopic analysis of Jupiter's atmosphere revealed vast amounts of methane—a carbon atom with four hydrogen atoms. The skies over the newborn Earth, Oparin said, weren't dominated by oxygen and nitrogen, as they are today—ours is a planet on which life has profoundly shaped the composition of the atmosphere—rather, the Earth's natal atmosphere consisted of methane, ammonia, water, and hydrogen. Here was the potential for a completely different kind of planetary chemistry than that which occurs on Earth today. It was from these fundamental molecular building blocks, Oparin argued, that life evolved.

Oparin suggested a long and detailed series of chemical reactions that, from these atmospheric molecular building blocks, could lead to increasingly more complex molecules, the ones we are familiar with in our daily diets: fats, sugars, and proteins. For Oparin, life emerged as an energetic or metabolic system that evolved through a series of increasingly complex chemical reactions. Chemical reactions in the atmosphere would have created a hazy rain of organic molecules that embedded the seeds of life in the Earth's Archean oceans, and in this chemical cocktail, Oparin imagined, the first protocells evolved.

What made Oparin's ideas revolutionary is that through detailed exposition he'd given other scientists the sine qua non of a respectable theory: it could be tested. They couldn't re-create the early Earth, but they could try to re-create its conditions.

THE EARTH IN GLASS

On a muggy August morning in 1957, more than three decades after Oparin's pamphlet first hit the streets of Moscow, and just as B^2FH—the famous paper describing the origin of the elements in stars—was going to print, Stanley Miller stood outside the Soviet Academy of Sciences, wondering at his good fortune to be a young American in the Soviet capital at the height of the Cold War. The summer heat only added to his mixture of excitement and apprehension. He knew that in the Red-baiting atmosphere at home, his trip to the Soviet Union could come back to shred his career. Yet any fear of the government men who'd visited him before his departure, urging him to keep his eyes and ears open, was ultimately fodder for discreet mockery rather than concern. After all, he was a chemist, interested in molecules, not a nuclear physicist with an eye for missiles. He wasn't concerned, as were the government agents, that the Soviets might be on the brink of creating artificial life. At the moment, what was more intimidating for this clean-shaven young chemist was that he spoke only English—he was out of his league amid a European polyglot ensemble of the world's leading biochemists, a group in which conversations switched rapidly from German to French to Russian. And he'd been studying biochemistry for only several years. But he wouldn't have missed being here for the world. Oparin himself had invited the young man to this historic event, the first International Symposium on the Origin of Life on Earth. Though they'd never met, Oparin and Miller already shared a special scientific bond, the bond between the theoretician and the experimentalist who proves him right.

Miller knew that his path to Moscow was a serendipitous one. In the spring of 1951, he'd arrived at the University of Chicago as a graduate student attracted by the university's "can-do" spirit. A decade before, in a basement lab on the university's village-like campus, physicist Enrico Fermi had sparked the first controlled nuclear reaction. Miller had quickly landed a plum position with Edward Teller, the scientist who'd be remembered as the father of the hydrogen bomb, to work on big-bang nucleosynthesis—how the elements might have formed at the birth of the universe. However, this astrophysics research took a U-turn toward biology during the fall of Miller's first year, when he attended one of planetary-science pioneer Harold Urey's famously inspiring lectures.

Urey was part of a generation of quantum physicists and atomic-bomb makers who, either inspired by the relationship between the quantum atom and life or disenchanted by the atomic bomb and its impact on lives, had turned from physics to questions of biology. He was perfectly positioned for the transition. He'd begun his career as a physical chemist working with Niels Bohr in Copenhagen, exploring the relationship between atoms, their isotopes, and the resulting chemistry of their interactions. The work led to his discovery of deuterium, a heavy isotope of hydrogen, for which he won the 1934 Nobel Prize in Chemistry. During World War II, as part of the Manhattan Project, Urey developed the critical technique for separating the isotopes of uranium to produce bomb-grade material. Based on his knowledge of differences in isotope ratios, he began thinking about what these revealed about the story of the Solar System, particularly about the Earth's origins—the new field of cosmochemistry. He became one of the first to view the Earth as an evolving planetary geochemical system, rather than as a static, isolated orb.

Urey came to the same conclusion that Alexander Oparin had come to thirty years before: the newborn Earth, and the environment in which life first evolved, was nothing like it is today. Urey asserted that if researchers wanted to understand first life, they needed to get the primordial planetary chemistry right. Based on the planetary evidence, the newborn Earth's atmosphere, still steaming from its formation from a stellar cloud, was dominated by methane, ammonia, water, and hydrogen—the simplest forms of carbon, nitrogen, and oxygen fully bonded with hydrogen.

Looking out at a University of Chicago lecture hall of young students one morning in 1951, Urey suggested "that experimentation on the production of organic compounds from water and methane . . . and the possible effects of electric discharges on the reactions [simulating] electric storms . . . would be most profitable." For Stanley Miller, the words were electrifying, sparking a creative vision that caught even Urey off guard. At first, Urey had balked at accepting Miller as his graduate student. But Miller persisted, and Urey's abundant curiosity won out over any administrative reserve.

Now, six years later, it was Miller who stood before an audience at the Soviet Academy of Sciences. Everyone in the Moscow audience knew the details of what Miller would say—it was a chance for the crowd to celebrate the victory and linger on what had been achieved. Oparin himself had learned of the results in 1953, when a colleague had run up to him waving a newspaper, showing the Russian that what he'd proposed thirty years before had come to pass. "I don't believe it!" Oparin exclaimed. In fact, one of the most important biochemistry experiments of the twentieth century had been quite simple. What was more remarkable was that no one had tried it before, as it used glassware and equipment available in any university chemistry lab. Standing before his distinguished audience of scientists, Miller described how he'd created a bare-bones equivalent of the primordial Earth's atmosphere and oceans in laboratory glassware. He noted that others thought they had tried this, using various mixtures of water, carbon dioxide, and oxygen—molecules common in Earth's atmosphere today—but that they'd had disappointing results. What made Miller's experiment transformational was that he'd fully embraced Urey's and Oparin's Stardust Revolution thinking and had re-created not some version of today's atmosphere but rather what was thought to be the Earth's primordial atmosphere.

Miller's experimental apparatus consisted of two globe-shaped flasks, one slightly to the side and above the other, like two party balloons connected by a glass tube. The slightly larger and higher one stood in for the Earth's early atmosphere; the lower, smaller one for the planet's primordial sea. After sterilizing the entire system for a day, Miller voided the top flask of Chicago's oxygen-rich air and filled it with Urey's proposed mixture of methane, ammonia, and hydrogen. He partially filled the lower flask with water, which, when gently heated, fed water vapor into the mix of primordial gases in the higher flask. Then Miller flipped the electricity switch, and sparks jumped between two tungsten electrodes inserted into the gas mix. Here in glassware was a vision of an Earth lost to time: water, heated by the planet's fiery birth, evaporating from steamy early seas and mixing in the hazy atmosphere where one could imagine nighttime bolts of lightning illuminating a strange, wet, nascent world. Within days, Miller knew he'd done something special, and within a week, he knew it was dramatic. His Earth-in-glass had been transformed. The stand-in sea had gone brown; the walls of the upper chamber were coated with an oily sludge. Miller got to work analyzing the results, and what he found astounded him. His brown sea was a primordial soup of life. Amid the brown water and oily sludge, he separated out five amino acids—the building blocks of proteins. Among them were glycine, aspartic acid, and alanine, three amino acids that Miller knew were in him and in most living things.

He didn't even know how well he'd done. Months after Miller's death in May 2007, Jeffrey Bada, his former graduate student turned colleague at the University of California–San Diego, opened a dusty cardboard box he'd inherited from Miller. It contained hundreds of vials with the dried amino-acid residues from Miller's classic primitive-Earth-in-glass experiments. Bada arranged for these samples to be reanalyzed using modern, and far more sensitive, chemical-analysis techniques than the liquid-paper chromatography that Miller had used more than half a century before. With this keener analysis, nine additional amino acids emerged amid the residue of Miller's original experiment, and even more were discovered in later modified versions of the experiment.

Stanley Miller's results were the twentieth-century version of Friedrich Wöhler's creation of laboratory urea. Miller and Urey had added an enormous plank to the bridge between the chemistry of what scientists viewed as life and nonlife. As the *New York Times* reported, Miller "made chemical history by taking the first step that may lead a century hence to the creation of something chemically like beefsteak or white of egg." For those pondering the origins of life, however, these amino acids didn't set their sights on the future but on the distant molecular past.

Miller's results landed in an already bubbling biological pot of change. They were published in May 1953, less than a month after the momentous publication by James Watson and Francis Crick revealing the structure of DNA, the long-sought-after genetic molecule. Suddenly the question of the molecular origin of life was so much more than conjecture. Here now were the molecules of inheritance, DNA, and its handmaiden, amino acids, the molecules that join to form the proteins that are the core structural molecules of life. It was the dawn of a new kind of biology, molecular biology, in which the nature of life wasn't focused on species of animals but on the interplay of molecules—molecules that biologists now knew could form, in many cases, from the cosmos' basic building blocks.

This didn't mean that Miller had shown how life had actually started, though certainly some were drawn to this Promethean promise; that in the hyperbole of the Cold War, somehow they'd nailed the greatest scientific problem on their first try. After a presentation by Miller of his results to University of Chicago colleagues, Harold Urey famously quipped to Enrico Fermi that "if God didn't do it this way, he overlooked a good bet!"

Today, the exact nature of the Earth's first atmosphere—and the rate at which it changed as a result of volcanic discharges, atmospheric chemistry, the impact of meteorites and comets, and the eventual rise of life—is a hotly debated topic. Most geoscientists don't think the primitive atmosphere was as reducing, or as hydrogen rich, as the mix of chemicals in Miller's glassware. Instead, views range widely from a mildly reducing atmosphere—with greater amounts of nitrogen, water, and carbon monoxide—to one that was relatively oxidative—with yet greater amounts of water, carbon dioxide, and sulfur dioxide. Similarly, the locale of interest for protean chemistry had moved from the atmosphere to volcanic plumes and the ocean bottom, particularly the chemistry around "hot smokers," the mineral-rich hot vents on the ocean floor. What Miller's experiment did clearly show was the ability to gather the knowledge and skills to imagine the Earth as it was four and a half billion years ago and then re-create, in the most rudimentary form, the ingredients and conditions present when there was no life, only an organically rich planet, pregnant with promise.

Just months after Miller presented in Moscow, his results and Oparin's vision were about to take off into a whole new dimension: the space age.

LIFTOFF FOR EXOBIOLOGY

Like thousands of others in Calcutta, India, on the night of November 6, 1957, John Haldane and Joshua Lederberg occasionally glanced at the full Moon in anticipation of its imminent disappearance. Calcutta's stifling summer heat was subsiding, and the two men shared a rooftop dinner in the relatively cool evening air, waiting for the lunar eclipse. While others in the city talked about what the eclipse might portend for good or evil, few—if any—talked about the new scientific age that Lederberg foresaw. The thirty-one-year-old American microbial geneticist had come to visit one of his heroes, the British geneticist J. B. S. Haldane, who at the age of sixty-five had recently moved from London to Calcutta. They'd both been disappointed to miss the Moscow Origins of Life conference; Haldane, because of his move to Calcutta; Lederberg, because of work in Australia.

Lederberg, a New York rabbi's son, turned several generations of zeal for Talmud study into a precocious passion for the reproductive workings of bacteria. In 1946, at the age of twenty-one, he broke biological orthodoxy by showing that bacteria were not only far more genetically complex than previously thought, but that a type of bacterial sex—gene swapping—was going on, a discovery that won him the 1958 Nobel Prize in Physiology or Medicine. Lederberg was a rare breed of scientist with a deeply visceral sense of his field, one who integrated new events into a boiling pot of ideas and who thought deeply about their concrete consequences. For his part, Haldane was second only to Oparin as a father of twentieth-century reflection on the origin of life. In the 1920s, Haldane had been deeply involved in the development of evolutionary genetics, which, coupled with the discovery of viruses—remarkable microscopic organisms that appeared half-alive and able to survive only when infecting a host cell—set Haldane to thinking about evolution that might have occurred before the first cell, or even the first virus.

In 1929, five years after Oparin's initial pamphlet version of *The Origin of Life* appeared on Moscow's streets in Russian, Haldane independently, and without access to Oparin's work, published his own doppelgänger English version, also titled *The Origin of Life*. Both scientists were riding the same wave of insight: that the primordial conditions of the early Earth atmosphere had been vastly different from those of today and had been the site of organic synthesis that turned the Earth's early oceans into a primordial soup of life's building blocks. As a result, there wasn't an impassable abyss between the complexities of the cell and the relative simplicity of the foundational organic molecules; self-reproducing organic molecules might be an intermediate link between seemingly inanimate matter and life.

As Lederberg and Haldane talked intensely that night, the Moon became a source of great expectations and deep concern. For millennia, the Moon had been a harbinger of Earthly events, and now the question was not just what the Moon could do to the Earth but what Earthlings might do to the Moon. Just a month earlier, on October 4 (a month after Oparin's meeting), the Soviet Union had launched the world's first satellite, *Sputnik 1*, and with it, the space age. Lederberg and Haldane, both perennially at the forefront of their respective disciplines, were awed by this technological leap and the prospect of where it might lead. Their conversation turned to a disturbing possibility. What if, to mark the upcoming fortieth anniversary of the Bolshevik Revolution and make the ultimate display of power, the Soviets were to detonate a nuclear bomb on the Moon—to symbolically put a Red Star on the Moon? The idea would have seemed ludicrous several years earlier, but under the Calcutta Moon, it appeared completely possible. Two years earlier, the Soviets had detonated their first megaton-scale hydrogen bomb. The past summer, Soviet rocketeers had launched their first intercontinental missile, blasting it an impressive and terrifying four thousand miles. Now, humans really could reach out and touch what had hitherto always been out of reach.

To Lederberg, the once faraway and pure Moon appeared deeply vulnerable. Not just its cratered face but also potential lunar microbes that he believed could hold the key to understanding the origin of life on Earth and possibly beyond. Yet, before rockets were used to explore the Moon, the real prospect existed of contaminating it with nuclear waste. As a microbiologist, Lederberg knew how easy it was to contaminate a sample and destroy an experiment; he was also aware of the ubiquitous and exponential growth of bacterial populations. For Lederberg, this wasn't just late-night, alcohol-fueled speculation to be forgotten the next morning. More than any other biologist, he saw rockets and satellites not just as announcing the dawn of the space age but also as introducing a new age of biology. Here was an amazing and unique opportunity to explore another celestial body for life, an opportunity that could be ruined by national grandstanding or ignorance. The answer to Darwin's gap might lie within reach on the lunar surface, but humanity might spoil its once-in-a-civilization opportunity.

Returning to the United States, Lederberg began a determined one-man campaign to raise concern about lunar contamination, whether by radioactivity from a superpower nuclear stunt or by inadvertent bacterial contamination from a lunar rover. It was a campaign that would mature into NASA's planetary protection program. "Since the sending of rockets to crash on the moon's surface is within the grasp of present technique," Lederberg wrote in a paper published by the journal *Science* in June 1958, "while the retrieval of samples is not, we are in the awkward situation of being able to spoil certain possibilities for scientific investigation for a considerable interval before we can constructively realize them." Politicians, bureaucrats, and scientists listened to Lederberg, even more so after the autumn of 1958, when his Nobel Prize was announced. One of the upshots of Lederberg's campaign was the US Department of Defense Project A119, launched the same year, in which a young astronomer named Carl Sagan was hired to make calculations of the results of a lunar nuclear blast.

Also in that year, President Dwight D. Eisenhower officially entered the United States into the space race through the National Aeronautics and Space Act, creating NASA as the US space agency. Soon after, NASA's first deputy administrator, inspired by Lederberg, asked the National Academy of Sciences to set up a Space Sciences Board to advise NASA. The Space Sciences Board was divided into several subpanels, including one on extraterrestrial life, the forerunner of the Committee on the Origins and Evolution of Life. Its first chairman: Joshua Lederberg. But as Lederberg immersed himself in astronomy papers and talked with the committee members he recruited, including Harold Urey and Carl Sagan, he saw that what they were discussing was something much bigger than preventing lunar contamination. The space age was also liftoff for a new science. Lederberg dubbed it "exobiology": biology beyond Earth. "Twenty-five centuries of scientific astronomy have widened the horizons of the physical world, and the casual place of the planet Earth in the expanding universe is a central theme in our modern scientific culture," he wrote, in what appears as a precocious précis of the Stardust Revolution.

> The dynamics of celestial bodies, as observed from the earth, is the richest inspiration for the generalization of our concepts of mass and energy throughout the universe. The spectra of the stars likewise testify to the universality of our concepts in chemistry. But biology has lacked tools for such extension, and "life" until now has meant only terrestrial life. . . . For the most part, biological science has been the rationalization of particular facts, and we have had all too limited a basis for the construction and testing of meaningful axioms to support a theory of life.

Exobiology could address Darwin's gap by providing comparisons not only by studying life on Earth but also, Lederberg envisioned, by comparing terrestrial molecules with molecules on the Moon. Four years earlier, President John F. Kennedy had stood in Houston's Rice Stadium and delivered his famous "We choose to go to the moon" speech. Now Joshua Lederberg was envisioning that day and what it would mean, not for engineering, geopolitics, and the history of human adventure, but for understanding our origins.

The seeds of something profound had been planted in his mind that Calcutta night. Lederberg, the biologist, saw the Moon not as distant and other but as sharing a common lineage, one that might be preserved on the lunar surface. He thought that the Moon's ancient cratered surface might contain prebiotic molecules untrammeled by the geological processes that have transformed the Earth. In his article "Moondust," Lederberg sounded more like an astronomer, referring to a new upstart field that was challenging the way astronomers viewed the heavens. Lederberg argued that the keys to understanding the origins of life on Earth might be found in the microscopic pores and surfaces of moondust. The Apollo missions might be the equivalent of Darwin's journey on the HMS *Beagle*, providing vistas for contrast and comparison.

But it wouldn't so much be dust from the Moon that would deepen our understanding of our origins. Rather, it would be something that, after millennia of stargazing, astronomers had only just begun to glimpse: cosmic dust. With all eyes on the Cold War race to the Moon, it was technology from an earlier war that would open astronomers' eyes to a whole new realm, a realm *between* the stars.

CHAPTER 5

DUST TO DIAMONDS

We now know that our origins lie in the dust of interstellar space, that our Earth and ourselves are condensates of the dark gaps between the stars, the same yawning expanses that are visible within the Milky Way on any clear night.

—James B. Kaler,
Cosmic Clouds: Birth, Death, and Recycling in the Galaxy, 1997

THE ORIGINAL DARK MATTER

For anyone pitching possible venues for a family holiday, Canada's Grasslands National Park might not sound at first like a must-see destination. In fact, when I made the suggestion to my wife, she asked, "Why would we go to somewhere called Grasslands? What are we going to see, grass?" She's far from alone in this opinion. Located in the extreme south of the province of Saskatchewan, just north of the Montana border, near the middle of the North American continent, Grasslands is one of Canada's newest and least-used national parks. And yes, there's a lot of grass. The park protects one of the few remaining stretches of North America's bald prairie. Most of the North American prairie has been paved, plowed into wheat and canola fields, or pastured into cattle range. But Grasslands National Park is semi-arid, receiving mere millimeters of rain a year more than a desert. It is land too marginal for even the most failure-hardened farmer.

I realized while camping at Grasslands that what makes it remarkable is that you go not to see things but to be seen. Not by other people but by the elements. Of course, there are lots of sights: the majestic flat-topped buttes with views across glacial-spillway-carved valleys, the spectacular flowering prickly pear cacti that appear as bursts of color amid the grasses, and the bison, their powerful humped shoulders silhouetted against endless distance. Most of all, though, in Grasslands National Park, there's the sky. Saskatchewan's vehicle license plate motto is Land of the Living Sky, a phrase that captures something as ephemeral, fleeting, and yet powerfully real as a towering thundercloud looming over the land on a scorching August afternoon. Here, on the flat, bald prairie, the sky swallows all. You watch menacing storms move in from the distant horizon; the weather forecast there for the seeing. Above all, it is the Sun, in all its blazing glory, that is omnipotent. There is nowhere to hide—no shade, no place where you can escape the Sun's gaze.

When the Sun sets, though, the sky turns from stunning to sublime. This is when Land of the Living Sky takes on new meaning. Grasslands National Park is a dark-sky preserve, part of a current global trend to protect not just threatened habitats and species of animals and plants but also the simplicity of the truly dark sky, free from human-caused illumination. In this sense, Grasslands is about as good as it gets in the easily car-accessible parts of North America—in the center of the park, you're about twenty miles from the closest porch light or streetlight.

In the early morning hours, I awake to look at the stars. I've seen many a deep, dark night sky, but this view engenders awe. Over the buttes and valleys, the Milky Way spills from horizon to horizon, a great slit of luminescent eye, like some celestial sea creature peering out from the darkness. I know I am looking edge-on into the heart of our galaxy, seeing the cumulative light of hundreds of millions of stars. It's so bright I could read a book by the light of these distant suns.

What's also clear in this deep dark of night is that, for all its stars, the Milky Way isn't all light. There are dark lanes. Not the gaps of darkness between the stars, but here and there great swaths of darkness, starless eddies in the galactic stream. Here, or at sea, or in those remote corners of our globe where the night sky is still wild, it's not just the stars that stand out; it's the darkness. What I peer into over Grasslands National Park is a crucial piece of the Stardust Revolution: the mystery of the original dark matter.

For astronomers, the term "dark matter" has long referred to something that appears to be out there—usually observed, by implication, but with no explanation in the existing astrophysics or cosmology. The search to explain today's dark matter, the sought-after and wonderfully named phantom weakly interacting massive particles (WIMPs), is inferred through the need for more matter to explain the gravitational behavior of galaxies. The dark voids of the Milky Way were astronomy's original dark matter. In them, we would discover an intimate link between darkness and light, the birth and death of stars, and our own dusty beginnings. The astrobiologist Joshua Lederberg had imagined moondust as holding the secrets of human origins. But we'd have to travel far beyond the Moon to find our stardust past.

These dark patches amid the stars have intrigued stargazers for millennia. The Inca of Peru, deeply reverent stargazers from mountain sites such as Machu Picchu, gave descriptive names to the dark constellations they saw in the Milky Way: a writhing patch of darkness was "the serpent"; a blob, "the toad"; a large patch, "the llama"; and, best of all, a rope-shaped extension of the llama was dubbed "the umbilicus of the llama." As European astronomers from the seventeenth century used telescopes to probe deeper into the heavens, they discovered not just stars but also a surprising menagerie of relatively dark, blurry patches of a variety of shapes and sizes—smaller versions of the Milky Way–scale dark swaths. There were also intermediate luminous objects—though not stars—that were dark, but not totally dark. Astronomers dubbed these stationary fuzzy objects *nebulae*, Latin for "clouds." The term captured both the similarity of their appearance to terrestrial clouds and the nebulous nature of early astronomers' understanding of them.

The most famous of these nebulae, the Orion Nebula (also known as M42 or Messier object 42), was first reported by Nicolas-Claude Fabri de Peiresc in 1611. In the late 1700s, the great British astronomer William Herschel described small, glowing clouds that he called planetary nebulae because of their circular shape (thus sowing seeds of confusion for future generations because these nebulae aren't related to planets but to dead stars), made from "a shining fluid of a nature totally unknown to us." By the early twentieth century, as some astronomers categorized star types, a smaller group turned to mapping the dark bits between the stars. In 1919, the American astronomer E. E. Barnard, after years of observing (while wearing his caribou-skin coat to stay warm), amassed a catalog of almost two hundred dark objects. Barnard was convinced that these dark clouds weren't just starless gaps of empty space but rather that they were some *thing*, some dark matter in the heavens obscuring starlight in these regions.

At about the same time, Roscoe Frank Sanford submitted his PhD thesis "On Some Relations of the Spiral Nebulae to the Milky Way," based on research he'd performed at the Lick Observatory in Northern California. A technically skilled astronomer, Sanford had carefully photographed dozens of distant, luminous pinwheel-shaped clouds, the enigmatic spiral nebulae, objects we know today as galaxies. What he found most remarkable was what he saw when he viewed these spirals edge-on: a dark, circuitous path that divided them in two. In a great leap of recognition, Sanford made the step from the *out there* to *right here*. He knew another such dark path amid the stars much closer to home—that of the Milky Way's dark clouds. At the time, "Milky Way" referred to the visible structure, whereas today it refers equally to the whole galaxy. That dark path, Sanford argued, implied that the spiral nebulae were in fact distant universes, similar to the Milky Way. Even more, it explained why we don't see these nebulae along the plane of the Milky Way: our view, Sanford concluded, is blocked by some obstructing matter, "whatever it might be."

The final missing piece of this original dark-matter puzzle would come half a dozen years later from one Robert Julius Trumpler. Trumpler is hardly a household name, but, like Hubble, he fundamentally changed astronomers' view of the cosmos. He showed that it's dusty—that when we look to the heavens, we're looking through the equivalent of a cosmic dust storm—and, in some places, it's more dense than others, but overall, dusty all the way down. Trumpler didn't look for cosmic dust but rather stumbled on it, as a kind of latter-day Columbus happening on a new world. Born in Zurich at the end of the nineteenth century, Trumpler as a young man turned from banking, the career his parents encouraged him to pursue, to one that appeared less lucrative but into which he could pour his curiosity and perfectionist's attention to detail: astronomy. His early research involved positional astronomy, mapping the locations and motions of the stars. His first paper used astropositioning to determine the exact latitude of his university town, Göttingen, Germany. It was this Zen for cosmic mapping that would lead Trumpler to new lands. In 1919, the Great War having come to an exhausted end, Trumpler began work at the Lick Observatory on the same telescope that Roscoe Sanford had used to spot the dark lanes in spiral nebulae. Soon Trumpler was mapping the distribution and measuring the size and distance to open-star clusters—spherical collections of several thousand stars gravitationally bound to one another and orbiting the Milky Way like fish in a school swimming the galactic currents.

Trumpler surveyed a hundred open-star clusters and found an odd result. The expectation was that clusters having the same number of stars, brightness, and light fingerprint would be about the same diameter across. But they weren't. Trumpler's measurements seemed to indicate that distant clusters were larger than nearby clusters of the same type. It was as if star clusters closer to home were dime-sized and that those farther away were silver-dollar-sized. This didn't make any sense. It would be returning to a pre-Copernican view of the heavens, in which the Earth was somehow a point of reference for structure in the cosmos, with star clusters increasing in size as they got farther from the Earth. Analyzing his results, Trumpler found another pattern: the light from the open-star clusters was blocked in a characteristic way with distance. Every 3,260 light-years, a star cluster's light dimmed by about two-thirds in every direction of the sky, except along the Milky Way's dark bands, where the dimming was more pronounced.

Trumpler also noticed that the light from more distant objects was not only diminished but also reddened. This reddening is different from the more famous red shift of starlight used by Hubble to calculate the speed of receding distant galaxies. In red shift, the entire spectral fingerprint of a star or a galaxy is shifted toward the red because the observed object is moving away from us. In what Trumpler observed, it wasn't that the spectrum shifted but rather that the interstellar material was blocking wavelengths at the bluer end of the spectrum while longer, redder wavelengths could pass through it. This is the same physical process that produces photogenic red sunsets. Trumpler realized that the way the light was filtered—what astronomers now call interstellar extinction—provided a wonderful clue as to what was blocking it. In order to redden the observed light, the size of the objects blocking the light must be about the size of the shorter blue wavelengths of light. There must be a fine haze between the stars, tiny bits of dust about the same size as particles of cigarette smoke or diesel exhaust.

In 1610, Galileo had used a telescope to see that the Milky Way is composed of countless stars, opening the way to a new understanding of the heavens. It had taken more than three and a half centuries for stargazers to see what was between those stars. Trumpler had discovered cosmic dust.

A NEW LAND BETWEEN THE STARS

Finding cosmic dust didn't grab headlines in the way that the discovery of an expanding universe did. Unlike Hubble, Trumpler isn't the moniker of a major telescope, yet the discovery of stuff between the stars was just as paradigm breaking as Hubble's discovery. Whereas Hubble added space by growth, Trumpler added space by increasing the available cosmic real estate. Trumpler had discovered more than dust; he'd discovered a vast unknown land, the interstellar medium. Space isn't empty. Even with all the advances of nineteenth-century astronomy and a deepening understanding of the stars, astronomers were still transfixed by the notion of the stars as points of light in a crystal-clear firmament. Even as late as the mid-1950s, there was a deep, powerful belief that space was empty, a realm as clean as polished black granite between the shining stars.

This age-old notion of interstellar emptiness held sway because astronomers found cosmic dust odd. Though for several centuries there'd been a sense that something was up with the dark bits, that idea didn't fit with our understanding of the cosmos. There was no cosmology or astrophysics to explain where this dust was coming from or going to. For example, as early as 1847, the astronomer Friedrich Struve had noted and calculated the rate of interstellar extinction of starlight, though he offered no mechanism to explain it. For more than a century, there'd been a form of dust denial, much like tired householders unwilling to face the dust bunnies growing in a home's corners. It was easier to alternately deny or ignore the dark bits in favor of the illuminating stories the stars had to tell.

By the 1950s, however, cosmic dust was working its way out of obscurity and into astrophysicists' thoughts, particularly into those of Fred Hoyle. In 1957, the same year that Hoyle copublished his paper on how the elements are forged in stars, he published another more speculative piece, his science-fiction novel *The Black Cloud*. The book tells the story of a massive cloud of interstellar matter, the enigmatic stuff between the stars. The cloud moves between the Sun and the Earth, blocking sunlight and threatening life on our planet. But a small group of smart Earthlings based at Cambridge University (where Hoyle happened to work at the time) learn to communicate with the cloud's alien intelligence and ask in that polite British way,

"If you don't mind too much, could you please move?" The Earth was saved.

In the novel's foreword, Hoyle wrote: "I hope that my scientific colleagues will enjoy this frolic. After all, there is very little here that could not conceivably happen." It was classic Hoyle: the appearance of self-deprecating humor followed by a "just try and challenge me" defense of his right to imagine. The fact was that most astronomers still viewed the real cosmic dark clouds simply as big, mysterious things that blocked light. Hoyle, and a handful of other astronomers, had a hunch that these dark bits were much more integral to cosmic ecology—and, somehow, that they were intimately connected with us.

In 1960, working with Chandra Wickramasinghe, a young post-graduate student from Ceylon (now Sri Lanka), Hoyle turned from the origin of the elements to the origin of the dust between the stars. Physicists use math to understand and visualize phenomena and objects they often never see, from the fusion reactions deep inside stars to the wave-particle duality of electrons. Among physicists, there are those for whom this understanding stays on the page—they can do the math, but the object always remains as if in a dream reality. A few, however, develop a visceral, intuitive sense of the mathematics: it begins to live and breathe. Arguably more than any other astrophysicist at the time, Fred Hoyle had such a sense of the lives of stars. He'd experienced his professional coming-of-age with his work in stellar nucleosynthesis stars and had come to know stars as dynamic creatures that were born, that went through developmental stages, and that eventually died. When it came to thinking about the origins of cosmic dust, Hoyle approached the question with what were now the first inklings of cosmic ecology thinking.

With Wickramasinghe, Hoyle asked two fundamental questions when thinking about cosmic dust: What materials in the cosmos are common enough to create enough dust to produce the level of observed light extinction; and where could this material come from? The answers lay in Hoyle's favorite stellar product: carbon. In a scientific article published in 1962, Hoyle and Wickramasinghe proposed that the vast reaches between the stars were filled primarily with flakes of graphite, a substance known on Earth primarily for its use as pencil lead.

Through a series of detailed astrophysical calculations, the two men showed that, in red giant stars—which are factories of carbon production—carbon could condense out as microscopic flakes of graphite in the star's cooler upper atmosphere. This was a radical, and yet in some respects obvious, idea. After all, where did all that carbon the star made go? Once formed, the star's outflowing radiation pressure—the force exerted by light waves—would expel the graphite into interstellar space at supersonic speeds of more than six hundred miles a second. The graphite would literally be fired out by light waves as tiny black dust ships carried on the stellar wind. Later, Hoyle and Wickramasinghe showed that the interstellar-dust spectroscopic fingerprint fit cleanly with that of pure graphite, as well as with the observed interstellar extinction.

Over the next decade, Wickramasinghe led the study of cosmic dust as Hoyle's colleague rather than as his student. Based on elemental abundances, the two researchers and others predicted that stardust would also contain large amounts of silicate dust—mixes of silicon and oxygen such as silicon dioxide, or quartz, the main ingredient in beach sand and glass—and that supernova would produce dust rich in iron and manganese.

Hoyle and Wickramasinghe had established a new way to think about dust—not as a hindrance to seeing but as an avenue to understanding cosmic processes. Stars weren't merely blocked by interstellar dust, they were making it. But that was only part of the dark-matter mystery. Where did the dust go? What was happening in those dark clouds? Truly understanding dust's cosmic role would require looking not through dust but at it, and this would require a whole new way of seeing.

SEEING WITH STARDUST EYES

When it comes to seeing the universe, our unaided eyes deceive us. We have eyes evolved for life on Earth. They are exquisitely tuned to spotting the movement of potential prey, reading text on a computer monitor, or enjoying the eruptive range of color in a perennial garden at its summer peak. Yet when it comes to seeing beyond our planetary home into our cosmic neighborhood, we've come to realize that with our eyes alone, we're largely blind. This might come as a shock to anyone who has ever looked at the night sky, and indeed it was a change in perspective that was vigorously resisted by many, if not most, twentieth-century astronomers. After all, we can see the stars—and that's what's out there, right?

This change in perspective can be linked to a single historical moment that took place in the year 1800. That was the year the astronomer Sir Frederick William Herschel decided, in a quirky bit of Age of Enlightenment experimentation, to measure the temperature of a rainbow. Anyone else wanting to measure such a thing might have been considered eccentric, but Herschel, who'd just turned sixty-two and thus was beyond caring what anyone else thought, was a consummate explorer of the heavens. He built telescopes with his sister Caroline and, later, with his son John. In 1781, he'd discovered Uranus, the first new planet found since antiquity. While Herschel is famous for his planetary discovery—one that, mispronounced, continues to elicit peals of laughter from schoolchildren two centuries later—he opened a much larger window on the universe, one that extends way beyond our Solar System.

Intrigued by his observation that the different-colored filters he used for observing the Sun appeared to let through different amounts of heat, Herschel decided to experiment and see if the colors of the Sun's spectrum did indeed have different temperatures. He set up a glass prism and used three glass thermometers, their bulbs blackened to better absorb heat, to measure the temperatures of the individual colors of the Sun's spectrum from violet to red. Lo and behold, there were differences: the temperature of each successive color increased from violet, the coolest; to red, the hottest. What happened next shows Herschel to be an exemplar of that marvel that characterizes our species: unfettered curiosity.

Noticing the pattern, Herschel became curious and placed a thermometer just a smidgen past the solar spectrum's red end, where there wasn't any visible light. He waited several minutes and checked the thermometer. It had the highest temperature of all. Herschel had discovered a new kind of light: infrared—literally, *below* red—radiation. More than that, he'd shown that there are forms of invisible light. And he showed that what he called "calorific rays" were waves just like visible light—they could be reflected and refracted. After millennia of human civilization, Herschel took the first small step into an invisible celestial realm, the expansive realm of light that we can't see with only our own eyes. He'd made the first step into the terra incognita of the electromagnetic spectrum, opening the window to remarkable new ways of seeing.

Today we know that what is colloquially called "light" represents just a little more than a sliver of the full spectrum of electromagnetic radiation—from the most powerful gamma rays to the longest radio waves. It's easy to take for granted two centuries of human exploration and discovery of the electromagnetic spectrum. But unlike the maps produced by Columbus, Magellan, and Cook, the electromagnetic spectrum is a map of an invisible world of waves. Nevertheless, it is omnipresent, and we've become completely dependent on it. On any given day, we can wake up to radio waves; defrost a bagel with microwaves; text a message to a friend on our smartphone using slightly longer microwaves; turn on the television with the blink of the handheld remote's infrared eye; slather on sunblock to protect us from the Sun's UV rays; and, if need be, get a lifesaving look into our bodies with x-rays.

All these waves and rays are various ways to describe the same phenomenon: light. Each of them travels as waves, all at the speed of light: 186,000 miles a second. What differentiates them are the twin characteristics of wavelength and energy. The shorter the length of the light wave, the more energetic it is. Billionth-of-a-meter-long x-rays will pierce skin and soft tissue, stopped only by denser bone, while meter-long radio waves pass through our bodies like large lolling waves under a boat on the ocean. Most importantly for astronomers, just as with visible light, every wavelength of light carries information. We now know that while humanity is tuned to see the visible, the universe is shining in all the colors, or wavelengths, of the electromagnetic spectrum. The Stardust Revolution has been built not just on bigger and better telescopes but, more importantly, on completely different ways of seeing. And when we look at the heavens with different eyes, we see a different universe.

THE COLD AND DIRTY COSMOS

Only a handful of humans have spent as much time looking at the cosmos with stardust eyes as Michael Werner, the cherubic-looking, good-natured, white-socks-and-sneakers, bike-in-his-office lead scientist for NASA's Spitzer Space Telescope. Werner has literally spent his adult life looking at the cosmos in the infrared. He was part of a generation of astronomers who came of age with infrared astronomy in the 1960s. In 1968—the year before others of his generation gathered a four-hour drive away, at Max Yasgur's farm in upstate New York, to sing the songs of the dawning of the Age of Aquarius—Werner was finishing up his PhD at Cornell University, thinking not about the stars but about what infrared eyes revealed about the gas and dust between them.

"For most astronomers, dust was just a nuisance because it was causing extinction," says Werner. "Until infrared [astronomy] started, there really wasn't that much information or interest in dust per se." Werner's thesis adviser at Cornell was Martin Harwit, a pioneer in infrared astronomy who'd worked with Fred Hoyle as a postdoctoral student. Harwit helped Werner get a postdoc position at the new Institute of Theoretical Astronomy established by Fred Hoyle at Cambridge University, where a new view of dust was dawning. The key to the mystery lay in the infrared, which opened a window for a new type of sky gazer, the cosmic-dust scientist.

During World War II, German scientists developed the first application of infrared sensors to see in the dark; this was the origin of today's night-vision combat technology. Using lead sulfide—a crystalline material that functions as a semiconductor—the German technology was the first to actually see in the infrared. The lead-sulfide detectors were photoconductors, which sensed the waves as light rather than as heat and produced an image in the same way that visible light waves do when striking the pixels in a digital camera. During the Cold War, the Americans and Soviets eagerly adapted and improved on this new way of seeing to spy on each other from the first satellites. These satellites could now literally see in the dark by sensing differences in temperature and, thus, in an object's infrared emissions.

While governments focused on seeing one another's secrets, a handful of astronomers wondered what infrared surprises the cosmos might hold. They realized, however, that they had a problem: much of the infrared radiation from space is absorbed by the Earth's atmosphere, except for several narrow windows of vision that provide a very limited infrared view. The molecules carbon dioxide, water, and methane—now well-known as greenhouse gases—absorb and trap not just heat (infrared) radiated from the Earth but also infrared radiation coming from space.

Cornell's Harwit led the effort in the United States to get infrared telescopes first high into the atmosphere and then into space in whatever way possible. This usually meant piggybacking on military or related high-altitude aircraft test-and-spy missions. In one case, Werner and Harwit used an infrared telescope—built by another infrared astronomy pioneer, Frank Low of the University of Arizona—that was jammed into an open door of a Learjet® during NASA's high-altitude testing of the plane. Harwit also negotiated space on a US military rocket and by doing so sent the first infrared telescope into space. Its view of the cosmos lasted only five minutes, but these five minutes firmly planted the seed of possibility.

In 1977, Werner began working on NASA's Shuttle Infrared Telescope Facility, SIRTF, a mouthful of an acronym renamed the Spitzer Space Telescope after its launch in 2003 in honor of Lyman Spitzer, the US astronomer who, in 1947, was one of the first to propose the idea of space-based telescopes. From his work with Harwit, Werner was fully aware of the enormous technical challenges and potential heartache of trying to see in the infrared from beyond Earth. He also knew that for all of Spitzer's challenges, the mission's success ultimately depended on one thing, which he called its prime directive: Spitzer had to be cold. Really cold. As much as it is a telescope, Spitzer is a sophisticated space-based extreme refrigerator. The Hubble Space Telescope's lesser-known space-based observational sibling, the Spitzer Space Telescope is among the coldest pieces of machinery ever created by humans. In the infrared, heat, as measured by temperature, is the equivalent of brightness. Thus, for an infrared telescope, the colder it is, the darker its view. A really dark-sky look in the infrared requires a really cold telescope. In essence, an infrared telescope can sense anything that's hotter than it is, including cosmic dust.

Even the Hubble Telescope's Earth orbit is too hot for Spitzer. Reflected and reradiated heat from the Earth gently warms orbiting satellites, so the Spitzer team arranged for its telescope to orbit the Sun, trailing in the Earth's orbital wake. When I visited Werner at the California Institute of Technology, home to the Spitzer Space Science Center and the nerve center for the Spitzer Space Telescope, the telescope was about ninety million miles from the Earth, at about the same distance from the Sun, with an invisible communications umbilicus transmitting home its marvelous images of the cosmos. When launched, Spitzer carried a tank of liquid helium that chilled its infrared detectors below the ambient temperature of space. The refrigeration unit worked similarly to a household fridge. However, rather than recirculating the helium-vapor coolant with a pump, the tiny amount of heat the Spitzer's electronics produced slowly boiled off the helium, dropping the sensors' temperature down to about –457°F. In comparison, liquid nitrogen—the substance that creates fantastic science-class demonstrations of crystalline bananas that shatter with a tap—boils at –320°F. For Spitzer, the ice cubes in your refrigerator are like intensely glowing charcoal embers.

At –457°F, the Spitzer and Werner got a deeper, darker view of the cosmos than anyone had ever achieved. This is critical, since the average temperature of dense dust and gas clouds ranges from –441°F to –414°F, whereas the more diffuse gas of the interstellar medium gets up to a balmy 1,340°F. In fact, for most objects in the universe—from you to planets to newborn stars and vast clouds of cosmic gas and dust—anything colder than 5,840.6°F is brightest in the infrared.

Deep-chilling the telescope dropped the level of background infrared brightness from the visible-light equivalent of high noon in the cloudless Sahara Desert to the deep darkness of a moonless night. This gave the Spitzer's extra-large, pizza-sized mirror—a little more than thirty-three inches in diameter—an infrared view of the heavens a million times darker than anywhere from the surface of the Earth. The Spitzer sees at wavelengths of 3 to 180 microns; a human hair is about 50 microns in diameter. What the Spitzer gives, quips Werner, is a view of the old, cold, and dirty cosmos. Old, because Spitzer sees ancient light red-shifted into the infrared. But it's the cold and dirty view that has helped transform astronomers' sense of the cosmos. With this infrared view, Spitzer and other infrared telescopes have revealed dust in a dramatic new light. Far from getting in the way of seeing, cosmic dust is a missing link in our connection with the cosmos.

THE DUSTY MISSING LINK

Stars don't shed only light and heat. They shed dust. In fact, in the new story of the Stardust Revolution, starlight is in some ways the secondary story. The light is the aftermath of stars' alchemical creation of the elements, and stardust is the first step between a star and you. Stardust is an evolutionary missing link, like finding the bones of an intermediary species between humans and an earlier primate. Before astronomers understood stardust, they could see that we and the stars were made of the same elements, and some astronomers dreamed that somehow we were deeply joined by this common language of the elements. But there was no mechanism, no pathway for understanding our link with the stars. Dust is that missing link. In stardust, we see how stars transform themselves into the beginnings of all we see around us. Where dust once obscured what appeared to be important, it now illuminates a new pathway to understanding.

It's a view illustrated in a historical retrospective by pioneering interstellar-dust researcher Mayo Greenberg, who titled one of his articles "In Dust We Trust." There is a truth in the dust billowing out from stars across interstellar space, sometimes raging as great cosmic dust storms until these infinitesimal grains of soot and sand pile up into massive sculptural dunes, which, with primordial gases and the eternals of time and gravity, will give birth to a new generation of stars. Dust is one of the ways that generations of stars communicate. Stars are born in dusty cocoons and, in dying, are transformed into dusty nebulae, the raw materials for a next generation of stars.

In his office, Michael Werner searches through the Spitzer's online archive of hundreds of images for one of his favorites. When he finds it, his computer monitor exudes a sculptural aura of green, red, and white. "I call this the Continents of Creation," he says, making a playful comparison to a famous Hubble Space Telescope image of a smaller section of the area, dubbed the Pillars of Creation. *Continents* is an appropriate word, for dust and gas are clearly seen as the structure of the universe. It's a celestial portrait of monumental scope. Here is a new land. After crossing an ocean of space and time, there, on the horizon, is not a little dust, a haze obscuring our view of something more important, but a vast glittering realm of dust and gas—majestic, mysterious, magnificent.

What's remarkable about these now-iconic infrared images—what's so odd in some ways that we struggle to comprehend what we see with our stardust eyes—is that where we're used to seeing points of light amid darkness, we're instead seeing a roiling landscape. There's a borderland of spires and canyons encircling a great ventricle-like central chamber, tens of light-years across. It appears as geography, as solid, because it's a stop-action snapshot. It is a vast cosmic pot of heated dust and gas. The tumultuous landscape gives the sense of process. There is both something here and something *happening* here. What's happening is one of the most astounding events in the cosmos: stars are being born. Astronomers call this cosmic territory in the constellation Cassiopeia, W5. It's like a hospital room number, for this is the image of a vast stellar birthing unit and nursery.

In looking at and through dust, infrared astronomy has turned the idea of the fixed, eternal star into that of the lives and ecology of stars, finally revealing the long-hidden secret of their birthplaces in dust. It's in this cold, dark realm that starlight first twinkles. That stars were still being born was itself a revolutionary concept. In 1941, the Spitzer Space Telescope's namesake, American astronomer Lyman Spitzer, made the first suggestion that stars aren't just eternal lights but were in fact being born in the interstellar matter. The comment, made in a scientific paper submitted to the *Astrophysical Journal*, was rejected by the anonymous referee as far too radical and speculative an idea, so Spitzer removed it from his paper.

But just after World War II, the Dutch American astronomer Bart Bok and his Harvard colleague Edith Reilly took Spitzer's idea of star formation from the condensation of interstellar matter—finally published after the war—one step further. They said they knew the probable location of these cradles of stellar creation: small, dense, dark clouds. Bok and Reilly documented a particular type of isolated, small, round, densely dark nebulae, today dubbed Bok globules. By astronomical standards, they're relatively puny, ranging in size from about ten thousand to thirty thousand times the distance of the Earth from the Sun, with masses of ten to one hundred times that of the Sun's.

But Bok and Reilly suggested that these dark clouds were gravitationally collapsing to form stars. Akin to caterpillar cocoons, these dark clouds were sites of transformation and rebirth; in this case, the birth of stars. It was a tantalizing prediction, we hadn't yet witnessed star birth because it had been an event hidden by a curtain of dust.

In the early 1960s, just as Western dads were entering terrestrial hospital birthing rooms, Robert Leighton, a physics professor at Caltech, was about to pull back the curtain on stellar birthplaces. Leighton believed he'd find something new by surveying the sky with the equivalent of new eyes: using new wavelengths. Today, this is orthodoxy; then, it was closer to heresy. As a result, Leighton's epochal all-sky infrared survey, called the Two-Micron Survey (because it looked at the two-micron wavelength), was a shoestring operation. While, across town in 1965, NASA engineers with deep budgets were preparing a rocket to go to the Moon and were using infrared to observe other planets in the Solar System, Leighton and his PhD student Gerry Neugebauer pieced together a telescope from war-surplus infrared detectors and a mirror that was partially built in Leighton's office. They tested their homemade infrared telescope in an alley behind Leighton's Caltech office, and when it passed this backyard test, they mounted it in a garage-like building on nearby Mount Wilson. The telescope was programmed to automatically scan the night sky, each night imaging a different strip of sky.

The results of the first all-sky infrared survey were electrifying. Leighton and Neugebauer's view of the heavens was like nothing anyone had ever seen. There, along with the timeless stars of the night sky, were hundreds of others that no one had seen before, not even with the most powerful optical telescopes. These newfound stars included numerous monstrous stars, larger than any then known, which we now know to be newborn giant stars still enshrouded in their natal cocoons of cosmic dust. Looking with stardust eyes, Leighton and Neugebauer had revealed not just the old stars but, amazingly, new ones.

Since then, infrared astronomy has been the key to telling dust's story in the birth and death of stars from the edge of time to today. Infrared telescopes have provided astronomers with a previously dust-blocked delivery-room view of star birth. In 1983, using the first space-based infrared telescope—the Infrared Astronomical Satellite (IRAS)—American and European astronomers peered into hundreds of dark Bok globules with the telescope's infrared eye and realized that Bart Bok was right. There, inside a quarter of the globules, was a baby star, and from this astronomers concluded that every Bok globule would eventually birth a star. Compared with Bok globules, larger dark clouds, such as Werner's Continents of Creation, aren't singular delivery rooms but the equivalent of a chaotic birthing floor in a major urban hospital. There are dozens of stars at various stages of birth, from those still wrapped in dust cocoons to meaty kids ready to go home, their energetic cries disturbing the entire ward. The stellar winds from the first stars to form blow away the surrounding dust and gas that accumulates at the edges of the winds, creating other areas of star formation.

The only stars not born in dust were the cosmos' very first stars. At the edge of time, just as there were no stars, there was no dust—not a single grain, anywhere. With the universe's first stars, not only did the cosmic lights come on, but with the fusion of metals, from carbon on up, came the cosmos' first dust. Spitzer has spotted galaxies with dusty light fingerprints dating back to only 870 million years after the big bang. The cosmos' first stars are thought to have been massive ones that burned bright and ended their short, hot lives as supernovas. Thus, the cosmos' first dust probably appeared as these stars burned, and it came from the supernova fallout as the exploding stars' ejecta gradually cooled and condensed. This dust was soon mixed with dust from carbon-spewing AGB (asymptotic giant branch) stars. This stardust was the cosmos' first solids, its first steps toward forming rocky planets.

As with the developmental stages of a child, this first dust changed the dynamic of all that followed. Astronomers believe that this primordial dust enabled the emergence of the first smaller, Sun-like stars—much smaller than the massive stars that die as supernovas. Smaller stars can form only when the collapsing cloud of gas and dust from which they take shape has a way to cool itself. For stars to become small, they first must cool, and dust, it turns out, is the best material the cosmos has to offer for cooling embryonic stars. As a stellar birth cloud gravitationally falls in on itself, the inward pressure heats the core, causing it to expand, as does any hot gas. Dust, however, absorbs the heat and reradiates it at infrared wavelengths that can escape the cloud, thus cooling the protostellar core and enabling it to further contract, allowing smaller stars to form. These stars don't require as much gravitational compaction to counteract heating-induced expansion. The first dust thus performed midwifery for a next generation of smaller stars, ones that in their dying days were copious dust producers.

From clean beginnings, the universe has become very dusty. Astronomers see cosmic dust everywhere. There's obviously dust between stars, but dust has also managed to work its way out of galaxies to mix with the copious primordial gas of the intergalactic medium. This said, it's important to keep dust in perspective. Even after thirteen billion years of stardust production, dust still makes up a very small percentage of the interstellar material. The stuff in the space between stars is about 99 percent gas—mostly the primordial mix of three-quarters hydrogen to about one-quarter helium, with only about 1 percent dust. While 1 percent dust doesn't sound like much, if the Earth's atmosphere had the same dust composition, you wouldn't be able to see your shoes.

The key to cosmic dust's importance is that it isn't evenly distributed. Just as dust drifts around a home, accumulating as more or less voluminous motes, cosmic dust also accumulates in quiescent corners of the cosmos. Interstellar space is remarkably windy; every star, to varying degrees depending on its type, produces stellar winds that are both motion and matter. These winds are a combination of light waves and the particles of dust and vaporized atoms that they propel. They drive cosmic dust and gas into cosmic-scale dust motes, from globules to vast dunes. This cosmic dust eventually begins to gravitationally collapse—to fall in on itself. When it does, the dark, cold nebulae—the places of star birth—appear.

Although stars are born from dust and gas, it is in their death throes that they return dust, old and new, to the cosmos in a great ritual of cosmic regeneration visible in their beautiful death shrouds, the spectacular planetary nebulae. About three hundred years ago, the light reached Earth from a massive star that had exploded as a supernova about eleven thousand light-years away, across the Milky Way. Today we call this dead star's brilliant, glimmering remnant cloud of dust and gas Cassiopeia A, the leading edges of which are racing outward at up to 3,750 miles per second—in other words, lapping the Earth in three seconds—with a shock-wave temperature of about fifty million degrees Fahrenheit, vaporizing everything it encounters. While Cassiopeia A is vaporizing dust at its leading edge, it has also made dust. Lots of it. According to data obtained from the Spitzer, at least ten thousand Earths' worth of new stardust produced by the exploding star cooled in the days and weeks after the explosion.

Supernovas, the deaths of giant stars, are relatively rare compared with Sun-like stars, and it's from these more common, modest stars that most stardust is produced. After Sun-like stars have burned all their hydrogen, they reach the equivalent of a midlife crisis, during which their cores collapse, becoming dense and hot enough to start burning helium as their primary fuel; this is the AGB stage. In 1969, the first infrared observations of these AGB-type stars found that they are shrouded in thick dust—sooty, sandy shells of graphite and silicates. In a similar fashion, the red supergiant star Betelgeuse—pronounced "Beetle Juice" and a star that, unlike our Sun, will end with a supernova bang—has become a prodigious dust factory in its dying days. On a clear night in the Northern Hemisphere, this notably red star, one of the brightest in the sky, shines at Orion's right shoulder. For all that shine, what's equally remarkable is the amount of dust that Betelgeuse is coughing out. Seen in the infrared, Betelgeuse is surrounded by a shroud of dust and gas, a vast glowing, clumpy shroud that extends out thirty-eight billion miles from the star's surface, about four hundred times the distance from the Sun to the Earth. Since the end of the last ice age, about ten thousand years ago, it has pumped out about our Sun's mass in dust, mostly silicates and aluminates—enough rock to eventually build hundreds of Earth-sized planets.

Thanks to infrared vision, cosmic dust no longer blocks light; it shines. In the infrared, it's possible to see dust for what it really is: microscopic minerals. The term *dust* conveys a sense of messiness and detritus. Carbonaceous stardust is often fluffy; the molecules bond to form powdery, spongelike masses. In this case, it's better named star*fluff*. But more often, it appears that the molecules that bond to form stardust do so in a highly ordered pattern that creates not the random-type dust that irritates homeowners but rather crystals—the elementary beginnings of every seaside pebble, farmer's field rock, or mountain cliff face on Earth, or anywhere else in the cosmos. With infrared eyes, we see that stars don't just produce light and heat, they also make rocks. The word *mineral* comes from the late fourteenth-century Latin term to describe something obtained from mining, dug out of the Earth. In the Stardust Revolution, we've extended the notion of mining and minerals to the stars, the field of astromineralogy. Infrared telescopes created the first generation of cosmic rock hounds.

Where once there was obscuring dust, there are now sparkling crystals. About a dozen astrominerals have been identified on the basis of their infrared light fingerprint, including graphite, moissanite (silicon carbide), and a green one called olivine. We know it's green because olivine is one of the most common minerals on Earth: it is found in the gemstone peridot, gives a green tint to some Hawaiian beaches, and is one of the dominant minerals in the Earth's core. Olivine is creating a green-hued sparkling crystal shower around newborn stars throughout the Milky Way, and it now appears that some of the crystals are the result of a kind of stellar blast furnace. Olivine is part of the silicate family of minerals, the most abundant minerals on Earth, a family that includes silicon dioxide, or silica, the main ingredient in glass. It forms at temperatures as hot as lava, and while this might occur in the shock wave of a supernova, the crystal might also be getting cooked up in stellar blasts from young stars that heat amorphous silicate dust, creating more organized crystalline forms. In before-and-after images from a prolonged period of stellar blasts, Spitzer scientists noticed a marked increase in the amount of crystalline material in the dust disk around a young star.

In the young field of astromineralogy, the one finding that has created perhaps the most buzz is the discovery of an astromineral that, here on Earth, is associated with the extremes of love, wealth, and strife: diamonds. The cosmos' first diamonds didn't form deep inside planets. They formed around stars. These symbols of eternal commitment were likely the universe's first crystals. At the very edge of time, glittering in the cosmos' first starlight, there were diamonds. On first glance, diamonds and graphite, or pencil lead, have nothing in common. But they're actually twins; both are pure carbon, just in different forms. Rearrange the carbon atoms, and graphite from a red giant star becomes astro-bling. These stellar diamond grains wouldn't impress most fiancées, at least not without a scanning electron microscope at hand. Stellar nano-diamonds aren't measured in carats but rather in carbon atoms; there are about a thousand atoms in a typical grain. Yet they are out there, possibly condensed from stellar outflows in a process analogous to the way artificial diamonds are made here on Earth; perhaps shocked into being by supernovas or maybe formed by the blast-furnace burps of newborn stars. As much as a third of the carbon dust from red giant stars emerges in the form of nano-diamonds. One way or another, the cosmos is a vast mine of them.

In one of the greatest about-faces in science, cosmic dust has gone from an astronomical nuisance to a critical missing link in our evolutionary journey from the stars. It's the great connector—we are indeed all connected through cosmic dust billions of years back and into eternity. While astronomers focused on the stars, it was in the cosmic dust that we'd find our gritty beginnings. The saying "dust to dust" starts with the stars. The bad news on the cleaning front is that you can't have our kind of universe without dust. Forget string theory, branes, and alternate universes. Without cosmic dust we wouldn't even have the universe we call home. Cosmic dust is essential for the birth of Sun-like stars, and it's the cosmic raw material for everything that we think of as "stuff," from meteorites to planets and from me to you. The dark bits are the universe's great element-recycling depots, the origins of stars and planets and, ultimately, people. A new breed of cosmic-dust scientists has shown that we're not just made of stardust; it's the ship for our collective epic cosmic odyssey.

Today in the Stardust Revolution, the talk about dust isn't focused on light extinction but on its role in the origin of life. In a 1962 article that started a new way of seeing—turning the haze between stars into stardust—Fred Hoyle and Chandra Wickramasinghe reflected on the cosmic ecological consequences of a universe full of tiny graphite flakes. They suggested that these tiny graphite grains might be the multitudinous equivalent of cosmic test tubes, ideal locales for interstellar chemistry and the formation of carbon-based molecules. Each grain of stardust might be a microscopic world, a tiny *terra nova*, a stage on which the next chapter of the Stardust Revolution could play out. But for this new view of stardust to develop required yet another new way of seeing.

CHAPTER 6

THE COSMOS GOES GREEN

The surface of the Earth is the shore of the cosmic ocean. From it we have learned most of what we know. Recently, we have waded a little out to sea, enough to dampen our toes or, at most, wet our ankles. The water seems inviting. The ocean calls. Some part of our being knows this is from where we came. We long to return.

—Carl Sagan, *Cosmos*, 1980

TUNING IN TO MOLECULES

In more ways than one, the fifty-mile drive from the University of Arizona campus in Tucson to the Kitt Peak National Observatory is a journey from one world to another. Mostly it's the journey from Starbucks to stardust. The city of Tucson is a modern-day desert mirage—a spreading metropolis of low-level, cacti-gardened, upscale homes in the middle of a dusty desert fed by dwindling supplies of water from the distant Colorado River. Driving out of the city, as the last strip malls, gas stations, and warehouses give way, you enter the world as it was—the Sonora Desert. Tall saguaro cacti, the walking cacti, appear to march up the low, rounded hills that undulate across the desert. It's an expansive, mysterious landscape, populated with exotic creatures such as scorpions; rattlesnakes; and the small, wild, very hairy pigs known as javelina.

Halfway to Kitt Peak, down Route 86 and about forty miles north of the Mexico–United States border, we pass a police checkpoint blocking the way. The police are searching for illegal aliens—Mexicans and other Latin Americans fleeing poverty and violence to enter this less-than-welcoming land of greater opportunity. The burly border-patrol guards, one holding back a sniffing German shepherd, wave us through. I'm traveling with two twentysomething astrochemistry graduate students from the University of Arizona's Steward Observatory. We're also searching for aliens, but of a different sort than what the guards and dogs are looking for.

While the US border patrol now tries to keep out illegal immigrants, early peoples in this area weren't too keen on the arrival of the first European settlers. The Tohono O'odham people have occupied this area for thousands of years. To them, Kitt Peak, its reddish rock rising two-thirds of a mile from the desert floor, a wonder of altitude amid this horizontal expanse, is still part of their tribal lands and is part of a sacred mountain range known as I'itoi's Garden. *I'itoi* is also known as *Elder Brother* or *Earth Maker*—the creator. To the Tohono O'odham, Baboquivari Peak, a massive cubic outcrop of granite to the southeast of Kitt Peak, is the center of creation, the place where all life began.

Standing atop Kitt Peak, you can believe it. In the cooler, greener mountain air, one feels a sense of otherworldliness looking out at an arid, barren landscape that appears to go on forever. For astronomers, Kitt Peak is also an ideal, scientifically sacred place to explore and contemplate the place where all life began. Situated high in the arid desert, the site has excellent "seeing" because the air column above it is often stable and dry. It also has excellent listening. The Kitt Peak National Observatory, which coordinates the cosmic viewing from the mountaintop, hosts a collection of more than twenty different telescopes, including the Arizona Radio Observatory's descriptively named 12 Meter Telescope. I've come to visit this "ear" to the cosmos.

The white dome housing the 12 Meter Telescope looks like the kind of shell that covers optical telescopes, but what's inside looks like a giant satellite dish. There's no long tube with glass mirrors for focusing light; "12 Meter" refers not to length but to the diameter of the massive metallic dish that acts as a great ear to the universe. It doesn't matter whether it's day or night: this radio telescope sees not light waves but radio waves—whispers from the universe. Just as an optical telescope focuses light, the large dish focuses incoming radio waves, and a receiver in turn amplifies the signal and channels it to computers and the astronomers inside the control room.

Just like the finely polished mirrors of an optical telescope, the 12 Meter's dish is a masterwork of fine-tuning. "There's only seventy-five microns of bumpiness across the telescope's twelve-meter face," says Bob Freund, the telescope's no-nonsense, long-time principal electrical engineer, as we stand looking up at the massive dish. That's less than half the width of a human hair and far smaller than the millimeter wavelengths at which the telescope "sees." If anything out there is broadcasting in the 12 Meter's range, the radio telescope has a good chance of hearing it.

If you're familiar with radio telescopes, it's probably because of their use in the search for extraterrestrial intelligence (SETI), particularly the giant, thousand-foot-wide Arecibo radio telescope in Puerto Rico. Perhaps you remember actress Jodie Foster, eyes wide with wonder, listening on headphones to an alien message via radio telescope in *Contact*, the movie based on Carl Sagan's 1985 book of the same name. The 12 Meter telescope is trying to make contact not with distant civilizations but with molecules. Radio telescopes have opened astronomers' eyes to another previously invisible part of the cosmos—its complex chemical nature.

To give me a sense of just how chemically alive the universe is, Bob Freund asks the controller to point the dish randomly at 60° north. The huge dish pivots until it's pointed up and out across the desert into the intense blue of the afternoon sky. We wait a moment as the telescope tunes in. And then, there it is on the display screen—the telltale peaks and troughs, like those on a cardiac monitor measuring a heartbeat, of the signal from trillions of interstellar molecules light-years away, vibrating to the energy from distant stars and sending their radio story out through the universe.

Later, as I drive with Freund back to Tucson, the setting Sun turning the desert into deepening shades of pink, he reflects on the marvel of his radio telescope on the mountain behind us. "It's just a big garbage-can lid," he confides to me. "With it, we can detect molecules light-years away. It's lousy with them out there." Fifty years ago, astronomers believed that it was impossible that even the simplest molecules could form around a star or in interstellar space. For these astronomers, the Earth was more than just the only planet with life; it was also the only known place in the cosmos with complex organic, or carbon-based, chemistry—the foundational molecules of life. But radio telescopes have revealed an otherwise invisible universe.

Radio astronomers haven't discovered alien life, but perhaps they have discovered something more profound. "Our observations suggest a universal prebiotic chemistry," says Jan Hollis, a veteran American radio astronomer who has discovered more than a dozen complex carbon molecules, including the first sugar found in interstellar space. Wherever astronomers look, from around dying stars to the depths of interstellar space—the cold, dense clouds from which stars emerge—they now see that the universe abounds with molecules. Far from being alien, most of these molecules look very familiar: they are the building blocks of life. Astronomers don't see other life, but they see life's precursor molecules, like footprints in the sand leading up to its doorway.

RADIO WHISPERS FROM THE UNIVERSE

The first person to actually hear cosmic radio signals thought he somehow had his wires crossed. It was a serendipitous discovery and without doubt the greatest scientific offshoot of the ongoing effort to avoid dropped phone calls. In 1927, the Bell Telephone Company in the United States introduced the first transatlantic call service. Seventy-five dollars bought callers three minutes of air time from New York to London via radiotelephone, which converted their voices into radio waves that were bounced from the ocean surface to the ionosphere, or upper atmosphere, across the Atlantic Ocean. But there was a problem: many calls were disrupted by electrical static.

Karl Jansky, a young engineer at Bell's Holmdel, New Jersey, laboratories, was asked to figure out what was causing the interference. Jansky embraced the project with gusto. He built a large antenna mounted on a turntable so that he could track the source of the interference and then doggedly began searching for it. By 1932, he'd figured out that thunderstorms were partly to blame. But there was something else that was producing a very steady hiss-type static. Jansky continued to track this hiss, turning his antenna this way and that, until he realized that the sound wasn't coming from anything on Earth; it came from above his head. He'd tuned into the Milky Way. His report, "Electrical Disturbances Apparently of Extraterrestrial Origin," marked the birth of radio astronomy.

For astronomers, the fact that the universe was sending out radio waves was greeted with head scratching and dismissal as a cosmic novelty rather than a serious topic. Astronomers had no experience with radio technology. Radio itself was new; radio broadcasts having begun only in the early 1920s. People felt awed by the ability to tune in to a new radio station across town; the idea of tuning in to the cosmos was just too far-out. Dutch American radio astronomy pioneer Grote Reber summed up most astronomers' responses to the discovery of stellar radio waves when he observed that they "could not dream up any rational way by which radio waves could be generated, and since they didn't know of a process, the whole affair was [considered by them] at best a mistake and at worst a hoax."

World War II changed that. The war transformed the way a generation of physicists and engineers, and some astronomers, thought about electromagnetic radiation. They received training in thinking not about waves of visible light but rather about microwaves and radio waves. During the war, legions of Allied and Nazi physicists and engineers raced to develop new and improved forms of radar, which gave them the ability to use radio waves and shorter microwaves to detect otherwise invisible objects that were obscured by distance and darkness. There was soon a clear hint that tracking enemy movements overlapped with astronomy. The most notable crossover occurred on the morning of February 12, 1942, when two German battleships passed through the English Channel undetected by British naval radar. The British were terrified. Had the Nazis developed a new radar-jamming technology? J. S. Hey, the British physicist assigned to troubleshoot the situation, discovered an unexpected, politically neutral source of the radio interference: the rising Sun. Somehow, the Sun's intense sunspot activity was producing a river of interfering radio waves.

After the war, astronomers at observatories across Europe, in Britain, and, to a lesser extent, in the United States scrounged up war-surplus radar materials. The most sought-after were the 7.5-meter Würzburg Riese antennae used in the German air defense radar system. Antennae that had been turned skyward to watch for Allied bombers were now turned to look deeper into the night sky, all the way to the stars. With all these antennae turned to the heavens, a new view—a radio view—of the cosmos began to take shape. It wasn't just the Milky Way's stars or the Sun that were communicating in radio waves, the universe was abuzz with them. In 1948, Cambridge astronomers using two Würzburg Riese antennae tuned in to an intense radio signal in the constellation Cassiopeia. They'd stumbled on the remains of a supernova, a stellar explosion still sizzling in radio waves. In 1955, two American astronomers trying to figure out what was causing interference with their radio telescope realized they were tuning in to Jupiter. By 1959, another pair of American scientists tweaked the idea that if you could tune in to stars and planets, you might also be able to pick up alien radio broadcasts from across the Milky Way. Within a year, American astronomer Frank Drake began the first radio-telescope search for interstellar communication, using the US National Radio Astronomy Observatory in Greenbank, West Virginia, setting the stage for today's SETI projects. By 1964, back at the Bell labs in New Jersey, Arno Penzias and Robert Wilson were trying to figure out what was causing the steady radio static on their twenty-foot radio telescope. At one point, they climbed up the antenna to clean it, thinking the static might be caused by encrusted pigeon poop. It wasn't the bird crap; it was the cosmos talking to them. At the same site where Jansky had been surprised by the song of the Milky Way, Penzias and Wilson had serendipitously tuned in to the cosmic microwave background, the attenuated birthing sounds of the big bang, a radio astronomy fluke that earned them the 1978 Nobel Prize in Physics.

For all this growing buzz over the sounds of the big bang or dying stars, it was on a much smaller level that a handful of astronomers believed their radio telescopes could make a huge difference. They didn't want to search for big objects but tiny ones: atoms and molecules. This molecular radio dreaming started in Nazi-occupied Holland during a gathering of Dutch astronomers in mid-1944, at the largely empty Leiden University where PhD student Hendrik van de Hulst shared a remarkable insight. He imagined that just as atoms and molecules in stars have distinct visible light fingerprints, they also emit distinct signatures at radio wavelengths—each molecule is its own tiny radio station, broadcasting at a particular frequency. So, van de Hulst suggested, if you knew a molecule's frequency, you could tune in and see if it was in outer space, as if tuning in to your favorite radio station.

Van de Hulst's quantum calculations indicated that every neutral hydrogen atom should emit an atomically faint radio signal, and since hydrogen is the cosmos' most abundant atom, this signal might turn into a hydrogen radio roar. And, van de Hulst said, he'd calculated the channel: look for it at a wavelength of twenty-one centimeters. By 1951, three research groups, using surplus war equipment, tuned in to interstellar hydrogen, soon revealing the Milky Way to be full of vast, diffuse clouds of it. This discovery of galactic clouds of hydrogen didn't, however, spark a radio search for other atoms or molecules. Hydrogen, astronomers believed, was in a class of its own—it was the abundant, simplest building block of the cosmos.

Hydrogen could survive, but there was no point in searching for scant amounts of other atoms, let alone molecules—the marriage of two or more atoms. The cosmos was simply too harsh a place—too hot or too energetic—for molecules either to form, or, in the off chance they did form, to avoid being immediately ripped asunder. The cosmos, they believed, was elemental.

This view dominated, even though there was some evidence to the contrary. In 1937, Mount Wilson astronomers had produced an unusual stellar light fingerprint using the observatory's new high-resolution spectrograph. The stellar light fingerprint had a single prominent line that caught the eyes of two spectroscopists in Belgium. The line is clearly produced using a Bunsen burner to heat not a single element but a molecule: carbon hydride. It's as simple a molecule as there is, a single carbon atom in an electron-sharing dance with a single atom of hydrogen. For astronomy, though, this little molecule was a giant leap forward—it was the first interstellar molecule.

Like radio waves, interstellar molecules weren't supposed to be there. More importantly, as with cosmic dust, astronomers had no way of incorporating them into their understanding of the cosmos. But by 1942, astronomers and chemists had identified two other molecules. The Canadian astronomer Andrew McKellar had spotted the light fingerprint of cyanogen—a carbon atom bound to a nitrogen one—in interstellar space; and in the scientific backwater of Saskatoon, Saskatchewan, the molecular spectroscopist Gerhard Herzberg—who'd barely escaped Nazi Germany, arriving in Saskatoon with only his wife, five dollars, and his spectroscopy equipment—identified another light fingerprint in the Mount Wilson spectra as that of the positive carbon hydride ion, or carbon and hydrogen minus one electron. Yet, rather than opening the door to exploration for more molecules, these discoveries reinforced most astronomers' view of an atomic cosmos. It appeared that only the most bare-boned two-atom molecules—a single carbon atom married to either a hydrogen or a nitrogen atom—existed in the dark interstellar void. It wasn't until December 1968, while all eyes were on the Apollo 8 lunar orbit, that a renegade radio telescope tuned in on a small molecule with one atom more. Just three atoms, but these three together make a universe of difference: water.

COSMIC WATER MAN

In the 1960s, NASA's Apollo Moon missions provided the first live television images of the Earth's closest cosmic neighbor, and it wasn't just sleepy-eyed kids in pajamas who were hoping for thc ultimate full-Moon revelation: signs of lunar life. With the first images, these ancient dreams shriveled, no more so than when it came to that defining ingredient of life, water. The sense was captured in the view of the lunar landing site from the *Eagle*, the Apollo 11 landing module. For all its watery allusion, the Sea of Tranquility appeared to half a billion television viewers as Buzz Aldrin described it, a "desolate beauty." Neil Armstrong left a first footstep that would never be erased by wave or rain. Any glimmers of hope for lunar life evaporated with that first small step by man on what appeared to be a bone-dry Moon.

This lunar view of a parched universe beyond Earth's borders was the follow-up punch to the first-ever close-up images from Mars, provided by NASA's 1964 Mariner Mars orbiter. Regardless of Mariner's seafaring moniker, the images the probe beamed back from the red planet revealed a barren, rock-strewn, and, above all, dry Martian surface. The images sank a century of sometimes jubilant speculation about a possible diverse Martian canal system and the civilization that must have built it. Mariner and other research also showed that even the Martian polar caps, long thought to hold vast amounts of water, were in fact mostly frozen carbon dioxide, with only trace amounts of icy water. The red planet appeared to be a dry and, more important, dead planet.

For those hoping that these inaugural journeys into our cosmic backyard would reveal hints of alien neighbors, the results were enormously disappointing. After all, NASA's mantra in the search for extraterrestrial life then as now is "Follow the water." Where there's life, there's liquid water. On Earth, this is a truism. Whether it's at the bottom of the world's deepest gold mines miles underground or in the eternal darkness at the bottom of the Pacific Ocean, wherever we've looked on our planet, if there's liquid water, there's usually life. Liquid doesn't mean an ocean, a lake, or even a puddle; liquid can be just a layer, no more than microns thick, a boundary layer of just less than solid ice, its melting caused by the friction as a mile-thick Greenland glacier moves over granite bedrock. Even in this unbelievably remote, extreme, and shifting environment, microbes spring into action, in what to them is the liquid of life as much as the Pacific is home to Tahitians.

Why is water so critical to life? We think of ourselves as solid beings, flesh and bone. But we're really made up of trillions of bags of water—our cells. When you step on the scale to weigh yourself, more than half of the answer, about 55 percent, is pure water. Water is the medium of our living interaction with the world. We pee out waste products, drink in dissolved nutrients, and breathe out water, a by-product of respiration, the burning of chemical energy that gives our cells the energy of life. It's why chemists call water the universal solvent. Life on Earth is immersed in, depends on, and originated in water.

As Earthlings watched successive Apollo astronauts bound across the desertlike lunar surface and let handfuls of dry lunar regolith—the Moon's powdery surface material—sift through their gloves, they silently received one clear message: in the search for extraterrestrial life, there appeared to be no watery tracks to follow. However, in Houston on that historic day in July 1969, there was one scientist who knew that the grainy black-and-white television images didn't tell the whole story. As the chairman of NASA's Apollo Science and Technology Advisory Committee, Charles Townes had a front-row seat for the lunar landing, seated beside Apollo project leader George Mueller and watching the touchdown on a large TV screen. Like the others, Townes could see that *Eagle* hadn't spied cosmic water, but he knew that he just had.

Charles Hard Townes was around that NASA table because, when it came to physics in the second half of the twentieth century in the United States, no one was more respected and accomplished than Professor Townes. He'd suggested the Apollo Science and Technology Advisory Committee several years earlier in his role as provost, or academic chief, of the Massachusetts Institute of Technology. Yet, for those around the NASA table, it wasn't only Townes's academic administrative achievements that gave him scientific street cred. As Neil Armstrong positioned a mirror on the lunar surface, it was thanks to Townes that this mirror would be used to measure the distance from the Earth to the Moon, by reflecting a laser beamed from Earth. Townes coinvented the laser and its predecessor, the maser—a molecular version of the laser—for which he shared the 1964 Nobel Prize in Physics.

For all his establishment credentials, Townes was a scientific rebel—of the nicest kind. He was a sought-after team player and could adroitly play that role to great end. But he was truly his happiest—blissful, really—when working alone or with a small team of passionate comrades at the misty, labyrinthine frontiers of the unknown. So, two years before the Apollo Moon landing, he'd quit his provost post, packed up his family, and moved from relatively staid Cambridge, Massachusetts, to the hotbed of 1960s activism: the University of California campus in Berkeley.

Some friends thought he was crazy. "Charlie, how could you move to Berkeley? That's the most sinful city in the whole United States!" chided his friend, the retired US admiral Chester Nimitz, namesake of the *Nimitz*-class aircraft carriers. While others in tie-dyed T-shirts and headbands were heading to the West Coast for anti–Vietnam War sit-ins and San Francisco's atmosphere of free love, it was another kind of freedom that drew Townes—the freedom to do whatever research he wanted. Another university suitor had tried to pin him down as to his research plan, but Townes balked. More liberal Berkeley offered him the no-holds-barred position of professor-at-large, including $100,000 to equip his lab and, given his

Nobel-laureate status, a coveted campus parking spot. Berkeley had one other, less obvious, thing that Townes wanted: a radio telescope. As tear gas wafted across the Berkeley commons, Townes had what most astronomers considered a radical and foolhardy idea: he wanted to search for molecules between the stars.

The likelihood of finding interstellar water was considered akin to the chance that Neil Armstrong would report back that the Moon was made of cheese. The universe—much like that view of the lunar regolith—was generally believed to be a dry, cosmic desert. In the two decades following the discovery of how stars create elements, one thing hadn't changed—astronomers still thought the universe was primarily *elemental.* While on Earth atoms join to form molecules—as occurs with water, carbon dioxide, and sugars, for example—astronomers believed that the rest of the universe was too diffuse and energetic a place for molecules to form and survive. We might be stardust, but we were fundamentally different from the rest of the cosmos. "We are complex; it is simple," went the thinking.

Townes, however, had come of age intellectually in the denouement of the great discoveries in quantum physics and the new era of its application. In the mid-1930s, he had earned his PhD at Caltech, where he talked physics on weekend hikes up Mount Wilson with the affable and brilliant young professor Robert Oppenheimer, who'd soon go on to lead the building of the first atomic bomb. When he graduated Caltech in 1939, Townes was disappointed that he couldn't get an academic post that would enable him to pursue his own research. He settled for an applied research job with the Bell labs, which turned out to be an ideal incubator for a man who would learn to talk with cosmic molecules. At the outbreak of World War II, Townes was assigned to the army of scientists tasked with developing new and better ways to see and therefore destroy the enemy. His job was to develop air force radar-guided bombing systems to enable pinpoint targeting, the equivalent of today's laser-guided smart bombs. As Townes and his colleagues developed one prototype after another, air force officials kept pushing them for more compact, lighter radar systems for planes. Townes worked on systems in the microwave region, first using a 10-centimeter wavelength, then 3 centimeters, and ultimately a measly 1.25 centimeters, about half the diameter of a US penny.

It was at this final level that Townes tuned in to something larger. He recalled that a decade earlier, a team of American researchers had discovered that when microwaves were beamed through ammonia gas, the molecules absorbed the microwave radiation right at the 1.25-centimeter wavelength. Radar designers were not concerned about this blocking their radar signals because there's very little ammonia in Earth's atmosphere. But there's lots of water in the atmosphere, and other radar researchers cautioned that water molecules were predicted to absorb at around the same wavelength.

For Townes, this caution wasn't so much a warning as a revelation. He saw that down in this largely unexplored netherworld of microwaves of around a centimeter in length was a whole new language, a way to communicate with molecules. Radar's sensitive microwave receivers could do more than lock on enemy targets; they could tune into otherwise invisible molecular targets. And just as radar could reveal information about a target—its speed, what it was made of—so microwaves could also reveal information about molecules, from their structures to their temperatures.

During his free time, Townes pursued the other side of his radio passion. Inspired by Karl Jansky's discovery of cosmic radio waves, he began pondering the physical atomic processes that might create them. After the war, first at the Bell labs and then as a professor at Columbia University, Townes set to work pioneering a new field, microwave molecular spectroscopy, investigating the invisible light fingerprints of molecules.

A molecule's unique light fingerprint, Townes and others realized, isn't caused by energy released and absorbed by electron quantum jumps or drops, as with single atoms, but because a free-floating molecule rotates, vibrates, or both. What's remarkable is that just as with the defined quantum jumps of electrons, this rotating and vibrating takes place only in discrete amounts, or quanta, at energies in the infrared and microwave. For example, Townes's favorite molecule, ammonia, is an amalgam of three hydrogen atoms in a pyramidal pattern around a central nitrogen atom. When the molecule is energized by a colliding particle or photon, it vibrates, the nitrogen going up and down like the surface of a trampoline, and the whole molecule repeatedly inverts itself like an umbrella turned endlessly inside out and back again—about *twenty-four billion* times a second. It's this vibrating, like the head of a molecular drum, that causes ammonia to emit a steady resonant beat at the microwave length of 1.25 centimeters.

Similarly, water's microwave signal is the result of one or more of three possible rotational movements the molecule makes, particularly one that emits a microwave signal at a wavelength of 1.35 centimeters, just a little down the microwave scale from ammonia. With microwave ovens we use this phenomenon in reverse, firing microwaves to get water molecules in potatoes, milk, or soup, for example, rotating en masse, hitting one another and everything around them, and in the process heating up our dinner.

By 1955, Townes and his Columbia University colleague and brother-in-law Arthur Schawlow had probed and prodded enough molecules with microwaves to write *Microwave Spectroscopy*, still the bible of the field. In this book, they outlined the theory, laboratory practices, and microwave molecular fingerprints that have guided subsequent generations of molecular spectroscopists. It was through these molecular insights that Townes envisaged the MASER (microwave amplification by stimulated emission of radiation), the first of which used the coordinated emission of microwaves by ammonia to create an energetic beam. It was just a jump along the electromagnetic spectrum for Townes, and others, to envision the laser, an optical light version of the maser.

While working on molecules in uptown Manhattan, Townes always had one eye turned heavenward and his ear cocked for celestial microwaves. At an astrophysics meeting in Washington, DC, in 1953, much of the buzz in the seminar rooms and during hallway coffee chats was about the 1951 discovery of neutral hydrogen by its radio fingerprint, as predicted by Hendrik van de Hulst during the war. Based on his molecular spectroscopy work, Townes, an astronomy outsider, presented a paper that Fred Hoyle still remembered years later. Townes argued that, if they were there in the space between the stars, it should be possible to detect a suite of simple molecules, including hydroxide (the marriage of a single oxygen and hydrogen atom), carbon monoxide, ammonia, and water. Few astronomers paid any attention to his prediction. Not only did they think it was a fool's errand, but they also didn't have the wherewithal to begin to look for these molecules even if they'd wanted to.

However, Townes infected one of his Columbia PhD students with enthusiasm for this stellar molecular mission. During the next half-dozen years, Alan Barrett risked his young career by searching for a molecule in space that everyone insisted wasn't there—until 1961, when Barrett and MIT's NASA-funded Lincoln lab engineer Sandy Weinreb detected a distinct cosmic microwave signal at eighteen centimeters, just the wavelength that Townes's group had predicted would be the cosmic molecular song of hydroxide. As the provost of MIT, Townes organized the celebratory press conference to announce the discovery. But while the cameras flashed and the journalists asked questions about this alien molecule that communicated in microwaves, Charles Townes was already moving on. His former student had doggedly pursued cosmic hydroxide and found it. Now Townes's own space mission was moving to launch.

Although Townes had free rein to do whatever he wanted when he arrived at Berkeley in 1967, that didn't mean he'd be encouraged in his search for cosmic molecules. "The chairman of the department of astronomy here, George Field—a very good theorist, an excellent scientist—he kept telling me, 'No, it can't be there. I can prove to you it can't be there,'" a ninety-four-year-old Townes told me in his chalkboard-dominated corner office on the top floor of Birge Hall at the University of California–Berkeley. By this time in his life, Townes was more than familiar with following his gut sense and his own eyes, when others thought that what he was doing was impossible. An example: on a walk in Copenhagen in the late 1940s, he'd described the idea for a maser to Niels Bohr. The revered father of quantum mechanics admonished his much younger colleague: "It's not possible, it's not possible." With a twinkle in his eyes, Townes says of Bohr: "I don't think he understood what I was trying to describe." Similarly, in 1945, Townes had gone to seek Ira Bowen's advice on using radio waves in astronomy. Bowen, Townes's former professor at Caltech and now head of the Mount Wilson Observatory, said: "Well, I'm very sorry to tell you, but I don't think radio waves are ever going to tell us anything about astronomy."

But at Berkeley, Townes found an ally in Jack Welch, an engineer in the university's electrical engineering and computer science department who was involved with the nascent radio astronomy lab. Welch was less interested in astronomy dogma and more interested in how astronomers might use the new Hat Creek Radio Observatory twenty-foot antenna he'd just helped install in a mountain meadow near Mount Lassen in Northern California. Its maiden voyage would be the search for molecules. The question was, which one first? Townes considered trying to tune in to carbon monoxide, but its dominant microwave signal is at just above two millimeters, barely wider than a piece of paper is thick, and the Hat Creek antenna's surface was too rough for the job—the two-millimeter signal, if there, would be distorted in the antenna's surface irregularities.

So Townes settled on trying to find an old friend: ammonia. With PhD student Al Cheung and postdoctoral fellow David Rank, he built an amplifier specially tuned to ammonia's 1.25-centimeter wavelength. In the early fall of 1968, Townes and his colleagues were set to go. They decided to begin their search in Sagittarius B, a mass of dark, dusty clouds near the center of our galaxy. As a reference point, they started by pointing their antenna toward the galactic center. Not a crackle. Then they moved it to point at Sagittarius B. From halfway across the Milky Way, in a region of space so different from Earth it's hard to imagine—a super-chilled –460°F—ammonia's microwave signal sprang to life in the Hat Creek Observatory's control room. Townes's 1.25-centimeter radar-guided bombing system had never flown, but, turned to the stars, he'd nailed his target on the first try. Cheung, Rank, and Townes had found interstellar ammonia. "How easy, and how exciting," Townes later recalled. Given the ease with which they had tuned in to ammonia, the energized group thought about their next target. It quickly came into focus just up the microwave dial, at 1.35 centimeters: water. With just one look, the trio came up wet in Sagittarius B.

Given the pace with which they were finding molecules, the question now was how widespread they were. Before announcing the water results, they decided to look elsewhere, including in the famous Orion Nebula. Townes was hosting a Christmas-week party for his research group when he got the call. It was from Al Cheung, whose PhD would be based on the work. The sound of a buoyant party made it difficult to hear, but Townes knew Cheung was up at Hat Creek, not letting the holiday get in the way of his search. "It must be raining on Orion," Cheung shouted. "It has a very strong water line." In the Berkeley foothills, water—in the form of champagne—was raised in celebration. They'd later calculate that the water line from Orion was twenty times stronger than that from Sagittarius B. More than discovering cosmic water, it was the cause of this powerful molecular signal, hitting them like blast from a fire hose, that would reverberate to Townes's core. "It turns out I didn't invent the maser," Townes tells me. "They have been out there for billions of years and nobody knew it."

Townes had the Nobel Prize, but when all was said and done, Nature had beat him to it. His team had found the first astronomical water maser, a natural maser formed when starlight excites vast clouds of water molecules, causing them to beam an intense, amplified microwave signal. Hundreds of water masers have since been discovered, some pumping out more energy than the Sun, all along the 1.35-centimeter wavelength, making them the most powerful molecular radio stations in the universe. For Townes, a lifelong Christian and in his later years a proponent of intelligent design, it was as if creation was winking at the boy from North Carolina, as if to say, *Yes, it had been here all the time, you just had to look.* Or, maybe, to say that it was only the beginning of what the search would reveal.

THE COSMIC SEA

The discovery of water beyond our Solar System unleashed a flood of research that has reshaped and continues to influence astronomers' view of the universe. Water—the popular marriage between the most common element to emerge from the big bang, hydrogen, and the most common element formed by stars, oxygen—is one of the three most abundant molecules in the universe, after molecular hydrogen and in an unclear photo finish with carbon monoxide. When we look into the dark depths of the night sky now, we know that the Earth, rather than a watery oasis, is a tiny blue drip in a cosmic sea.

Few single discoveries of the Stardust Revolution have so dramatically changed our view of the universe. In fifty years, we've gone from looking out at what we thought was a bone-dry cosmos to seeing one that's wet in every crevice and corner. Now knowing how to look, astronomers find water almost everywhere they search for it, the vast majority of it in the form of very cold gas or ice. The cosmos is dripping from the very edge of time to today; it's at times raining or snowing around newborn stars and dying ones; it's swirling in vast, gaseous torrents on the violent shores of massive black holes at the hearts of galaxies; it's icing over the dust grains in the cold, dark clouds that are stellar nurseries; and, in our Solar System, it's frozen into moon-sized ice balls around Saturn and Jupiter—even vaporized on the surface of the Sun. NASA's "follow the water" mantra in the search for alien life has taken on cosmic proportions.

Townes's team first heard cosmic water's radio splash, but astronomers found that it is also very visible in the infrared. Seeing water in space, however, requires getting above the Earth's water-vapor-laden atmosphere, which absorbs cosmic water's signal. It wasn't until the launch of the European Space Agency's Infrared Space Observatory (ISO) in 1995 that astronomers began to realize the extent of the cosmic sea. When ISO trained its infrared spectrometer on the Orion Nebula star-forming region where Townes had first detected cosmic water, the instrument revealed that what Al Cheung had called rain was in fact a deluge. Scientists at ISO determined that—over a region with a diameter about a hundred thousand times the Earth–Sun distance—the Orion Nebula region is producing enough water to fill Earth's oceans sixty times every day. That works out to twenty-one thousand Earth oceans a year. For millennia.

That's a lot of water, but it was only a hint of things to come. Cosmic water making in vast quantities has been going on for a very long time, and now astrophysicists are in a cosmic water fight to see who can find the oldest water. In 2008, a research team using a German radio telescope announced the detection of a water maser around a massive black hole whose light had traveled for 11.1 billion years, situating this watery signal 2.5 billion years after the big bang. This record took the plunge in 2011, when radio telescopes in Northern California and then Hawaii picked up the steamy hiss of water vapor from twelve billion years ago. Not only is this the oldest body of water ever found, it's the largest. The American astronomers estimated that the gaseous ocean they'd spotted spanned hundreds of light-years. And it held 140 trillion times the water in Earth's oceans.

Much closer to home, the news has been similar. From its status as the only water bearer in the Solar System fifty years ago, Earth has dropped to middling class. At least four other Solar System bodies contain more water. Jupiter's moons make the Earth seem puddle-like: Ganymede has thirty-six times as much water as the Earth, and Europa has almost three times as much. The water on the surface of these Jovian moons is all solid ice. However, most planetary scientists think there are deep liquid seas below these icy crusts. Similarly, recent lunar sampling has revealed that the once seemingly parched

Moon has abundant subsurface ice and thick water ice in craters at the lunar north and south poles—ice that could one day supply water for the inhabitants of a lunar colony. And, in the past decade, a series of NASA robotic missions to Mars has revealed that the red planet harbors abundant water ice reserves, and images of the Martian surface show rivulets possibly made by meltwater.

The presence of all this water has made astrophysicists turn to figuring out where it comes from. Few kids make it out of elementary school without drawing the Earth's water cycle—the endless movement of water evaporating from oceans; condensing in clouds; and falling back to the Earth's surface in rain, sleet, and snow to run back to the seas, where the cycle begins once again. Children of the Stardust Revolution can draw the cosmic water cycle. One big difference is that, whereas the terrestrial water cycle assumes the presence of water, the cosmic water cycle involves its formation from atomic scratch. And astrophysicists have discovered that, rather than cosmic water having a single wellspring, it is made in a myriad of ways. It appears that the cosmos pumps out water wherever it can. Whether the astrophysical environment is hot or cold, and whether stars are dying or being born, oxygen and hydrogen find each other and bond, sometimes for eternity.

There are at least four ways and environments in which hydrogen and oxygen meet and bond to form water. Just as older stars churn out dust and minerals, they also spout water. When the Herschel space telescope turned its sensitive infrared eye on the star CW Leonis in the constellation Leo, astronomers saw what was previously thought to be impossible. CW Leonis is a pulsating red giant star, also dubbed a carbon star for the amount of sooty carbon this elderly star emits. The astronomers expected to see lots of carbon, not steam. But there, deep in the star's atmosphere, was hot water vapor at a scalding 1,300°F. It had previously been thought that water wouldn't form around these carbon-rich stars because, before the oxygen could hook up with two hydrogen molecules, it would bind with the abundant carbon. Now it's believed that the water is formed when energetic rays of ultraviolet light from the star cleave molecules of carbon monoxide or silicon monoxide, liberating an atom of oxygen that then hooks up with the abundant hydrogen streaming by. It appears that CW Leonis is typical of stars between one and eight times the mass of the Sun: at the end of their lives, as they lose mass in the form of dust and molecules, much of what's poured out is water. In this way, the star both produces the oxygen and acts as the energetic matchmaker in catalyzing the formation of water.

And water is not just formed around dying or embryonic stars. The cold, dense molecular clouds in which stars are born also turn out to be ideal for making water. Water in turn repays the favor; like dust, it radiates heat in the infrared, enabling a collapsing cloud of dust and gas to continue collapsing. It's also thought to form in interstellar space, when molecular hydrogen is zapped and cleaved by a cosmic ray, the hydrogen then interacting with atomic oxygen.

But what really gets the scientists of the Stardust Revolution excited is the final place water is thought to both form and remain during its interstellar odyssey: on cosmic dust. During the advent of research on cosmic dust in the early 1960s, speculation arose from an analysis of their light fingerprints that these rocky or carbonaceous grains were iced over. Today we know that interstellar dust grains are in fact dirty molecular ice balls—bits of dust glued together with ice and encrusted with it. In interstellar space, it appears that dust plays the matchmaker in providing a surface that helps oxygen and hydrogen both meet and bond. In 2009, scientists reported that in a laboratory in France they'd pelted a mock interstellar grain with atomic beams of oxygen and hydrogen under interstellar conditions, and voilà, water had formed on the grain surface. Other researchers have found similar results and have discovered that grains ice over much faster than previously thought possible.

The combination of cosmic ice and dust might sound about as exciting as coming out of your house to find your car windshield covered with hard frost on a January morning, but for stardust scientists the combination of water and a solid surface is molecular paradise. Water was dubbed "the universal solvent" long before its cosmic origins were discovered. "Universal" pays tribute to water's ability to dissolve most things—to get molecules to mix and meet in its watery matrix, such as inside the human body. Water is life's matrix. Thus an icy interstellar grain becomes much more than just a dirty cosmic iceberg; it's the perfect place for something that was once thought to take place only on Earth: chemistry.

JOINING HEAVEN AND EARTH

Charles Townes's detection of cosmic water and ammonia involved more than the discovery of two molecules in outer space; it was the breaking of a cosmic molecular barrier. Water and ammonia aren't esoteric ionized molecules that can be dismissed as molecular space oddities; they are among the most familiar molecules on Earth, ones that until just then almost no one had thought could exist in interstellar space. And the number of atoms in cosmic molecules was marching upward. Water has three atoms; ammonia, four; what else was out there? Wherever they were, radio astronomers, and soon infrared astronomers as well, began tuning their telescopes for the messages from potential cosmic molecules. In the process, they transformed themselves into a new breed of astronomers: astrochemists.

When Townes pioneered the search for molecules in space, these scientists christened their nascent research niche "molecular astrophysics." A decade later, the field had matured from fanciful science fiction to peer-reviewed science—the transformative discipline of astrochemistry. The very term *astrochemistry* captures the philosophical phase shift that is the Stardust Revolution. Etymologically it's an oxymoron. Though the word's origins are debated, one possible source for *chemistry* derives from the ancient Egyptian word *keme* (*chem*), meaning "black earth"; it is also the name given to Egypt itself. By its very nature, chemistry is terrestrial; it's of and about the Earth. It relates to the very small, to atoms and molecules. In contrast, astronomy is about the nature of the stars, the largest bodies in the universe. With astronomy we look away from Earth, to the beyond; with chemistry we've conventionally looked inward at ourselves and the Earth. The new field of astrochemistry literally joins heaven and Earth as one system.

This transformation came just in time for Lucy Ziurys, now an astrochemist at the University of Arizona and director of the Arizona Radio Observatory's 12 Meter Telescope. "It was in 1977, [as] I began to write graduate school applications, that I read about the discovery of the first interstellar molecules. This just caught my imagination," recalls Ziurys, who entered Berkeley's astronomy department as a PhD student in 1978 to work in the new field that Townes and others were pioneering.

Ziurys is a child of the Stardust Revolution. Few have a more visceral sense of the grand hopes and sometimes false promises and dead ends that are the stuff of revolutionary science and technology. She grew up on the knee, literally, of the dream of space travel. Born on May 6, 1957, she was barely five months old when the Soviet Union's *Sputnik 1* satellite rocketed into space, the satellite's rhythmic and ethereal *beep, beep, beep* audible on shortwave from the Ziurys's Annapolis, Maryland, home. It was the Soviets' message to the world that the space age had begun. Eighteen months later, US president Dwight D. Eisenhower threw the Cold War space race into high gear with the creation of NASA, the National Aeronautics and Space Administration.

The Ziurys household was a space-age scientific incubator. Her father, Eugene Julius, was a member of the first generation of American rocket scientists. During the 1960s, Mr. Ziurys brought home stories of a world where the limits on the future were bound only by the imagination. After designing nuclear-powered commercial aircraft (which never, fortunately, took off), he became a project engineer with NERVA (Nuclear Engine for Rocket Vehicle Application), a program originated to use the power of the atom to blast humans into outer space. The program was supported by NASA and the even now futuristic-sounding Atomic Energy Commission Space Nuclear

Systems Office. Ziurys's last job before he retired was working on the development of nuclear fusion, the still-unfulfilled dream of producing energy on Earth as it's done in stars. Lucy Ziurys followed her father's space-science path, enrolling at Rice University in Houston, Texas, an informal preschool for NASA's nearby Johnson Space Center. But in her senior year, she faced the dilemma of the bright and deeply curious: she was equally fascinated by astronomy, physics, and chemistry—her mother Genevieve was an industrial chemist who taught high-school chemistry. At this point in their budding academic careers, most students specialize in one discipline or another. Ziurys didn't want to let go of any of these interests. That's when she stumbled across a quiet scientific revolution. For the young Ziurys, interstellar molecules weren't about a dawning scientific transformation but a personal one. Here was a field where she could search for new interstellar molecules using all the subjects she loved: physics, astronomy, *and* chemistry.

"The whole term 'astrochemist' used to be what astronomers considered a dirty word," says Ziurys. "When I had my grand idea of combining chemistry, astronomy, and physics, a lot of people laughed at me and said, 'Go into astronomy *or* work in the lab.' This idea of combining things wasn't very well accepted."

Now one of the world's leading astrochemists, Ziurys has a tomboyish pluck, a demeanor that speaks to the business at hand. During my visit at the end of a fall term, one of Ziurys's PhD students stood in her adviser's doorway late one Friday afternoon and said she was leaving for the graduate-student Christmas party. "What about the paper?" asked Ziurys, eyebrows raised, smileless. "My whole dream as a graduate student was to detect a new molecule," Ziurys says. "It never worked. So every time one of my graduate students finds a new molecule I just laugh. OK, you get to do what I didn't get to." Since her 1984 PhD thesis, she's more than made up for it. Ziurys has discovered, along with colleagues and graduate students, almost two dozen cosmic molecules, one-seventh of all those found—possibly the world-record number of such discoveries.

RED GIANTS AND WHITE DWARFS

Ziurys takes a two-pronged approach in probing the cosmos' molecular nature: she sees what can be cooked up in the lab, identifies its molecular spectrum, and then, with molecular light fingerprint in hand, she heads to Kitt Peak and other radio telescopes to see if the molecules are out there. Through this detailed mix of laboratory and radio telescope work, Lucy Ziurys is among a group that's been able to do more than spot interstellar molecules. She has helped lead astrochemistry to a new level: not just describing the detection of molecules but also revealing how and where they're formed. In the past decade, Ziurys and others have made a remarkable discovery: some stars make atoms on the inside and molecules on the outside. Just as Fred Hoyle, William Fowler, and the Burbidges described the process of how stars forge elements, Ziurys and other astrochemists are mapping out a spectacular, complicated process of stellar *molecular* synthesis. The same stars that fascinated Paul Merrill with their light fingerprints of heavy elements and Hoyle with their ability to make carbon are turning themselves into the molecules of life. "Organic chemistry starts back in the stars, where we have a carbon-rich environment," says Ziurys. Organic chemistry, the chemistry based on the carbon atom, starts not on Earth but around the very stars that have forged that carbon atom in their hearts: red giants.

There's hardly an epic Norse saga, whether of mighty Odin or, later, Beowulf and the monster Grendel, that compares with the true tale of our ancestors who in their dying days rose up as red giants only to end as white dwarfs. Over the past century, astronomers have gradually pieced together the life history of stars, like our Sun, that have a birth weight between one and eight times the Sun's mass. For most of their billions-of-years-long lives, from birth through middle age, these stars have stable lives—a period dubbed the main sequence—during which they're fueled by burning hydrogen in their cores. But in the last moments of their lives, over the course of less than a million years, and with much of the activity occurring in only ten thousand years, these dying stars shed about 80 percent of their mass.

This end-stage shrinking act begins when these stars have converted the bulk of their core hydrogen into helium. At this point, a layer of hydrogen begins burning around the star's helium core, marking the beginning of the red giant phase of the star's life. The "giant" part of the name comes from the fact that in these last moments the star begins to rapidly puff up, growing until an outer shell extends from ten thousand to as much as one hundred thousand times the star's original radius. With such a huge surface area, the star brightens enormously. As a result, the star's outer atmosphere is relatively cool and therefore appears red—hence, red giants. The best-known red giant star in the night sky is the bright star Mira, whose name, aptly, means "the wonderful."

In the star's next end-of-life stage, the asymptotic giant branch, helium in its core begins to fuse, or burn, transforming into carbon and oxygen. With each burst of core nuclear activity the star pulsates, a blast of solar wind carries the shell farther and farther away from the star's core. In the end, all that is left of the mighty star is a wizened white dwarf—a gravitationally trash-compacted carbon and oxygen stellar cinder, reduced to Earth-size, that continues to glow not from nuclear reactions but from its latent heat. About one in twenty stars is a white dwarf, and about 95 percent of stars are less than eight solar masses. Thus billions of stars in the Milky Way alone are going to lead geriatric red giant lives. Astronomers calculate that the Sun will take its turn as a red giant and end as a white dwarf in about 7.8 billion years.

You might well be wondering, where does all that disappearing star material go? The bulk of the star's body is transformed into molecules and dust. It turns out that in their death throes, red giants give the gift of life. They become what Lucy Ziurys calls "the most remarkable chemical laboratories in the universe." The vast majority of that chemistry is carbon, or organic, chemistry. If it seemed impossible to some theorists that stars might forge the elements, it was deemed equally impossible that stars were chemical factories. Until recently, astronomers thought that a star's outer shell, its circumstellar envelope, was a hellish, no-go zone in which all but the simplest and hardiest molecules, such as carbon monoxide, would be blown apart by the enormous temperature and the star's intense UV and particle radiation.

But what Ziurys has discovered is that a red giant's circumstellar envelope is more analogous to a chemical tidal zone, an area of enormous chemical diversity nurtured by its location at the boundary of "sea and shore"—the boundary of the dying star and the expanse of the interstellar medium. Using a radio telescope and infrared observations, Ziurys has constructed a chemical flowchart that maps the changing conditions in a red giant's circumstellar envelope. As in a tidal zone, it's characterized by enormous gradients that spur a cascading flurry of chemical reactions. At the inner edge of the envelope, the temperature is close to 1,300°F, and the density is a hundred thousand times higher than it is at the envelope's outer edge, where the temperature has plummeted to a frigid –414°F. Around just one red giant, IRC+10216, astrochemists have identified a smorgasbord of more than fifty different chemicals. About four in five are carbon-based, from methane to bulky eight-carbon molecules. There are also metal-containing molecules, including sodium chloride, or salt, and molecules containing magnesium and aluminum, elements forged through atomic synthesis in red giants.

As this dust and molecular soup expands outward, it forms bright planetary nebulae, with names like Cat's Eye and Helix Nebula—lit up like Christmas ornaments by intense ultraviolet radiation from the sizzling white dwarf. Gradually, over about ten thousand years, this cloud of stellar material dissipates, seeding the interstellar medium with a rich mix of organic molecules.

Since Townes's discovery, the past decades have been a period of astrochemical revelation. Where once the cosmos appeared as an atomic landscape, barren of all but the simplest chemical complexity, it has now emerged as molecular. The discovery of water initially made it one of a tidy half-dozen known cosmic molecules, but since 1970 the molecular stock index has been on a steady rise. By 1973, the astromolecular count hit 27 molecules; by 1987, it had more than doubled to 64; and by early 2012, it had reached about 140, with no end in sight. This group molecular soup includes mixes of at least fifteen different elements, from carbon, oxygen, and nitrogen to metals such as iron and aluminum, as well as those elements familiar from the label of any energy drink, including sodium, potassium, and phosphorous. Each of these molecules has been identified through a stringent process (with scientists looking over one another's shoulders and more than happy to prove a colleague wrong) that involves perfectly matching the hundreds of lines, and relative intensities, in each molecule's light fingerprint. The absence of only a single line rules out a possible detection.

When it comes to life's ingredients, the spiral arms of the Milky Way aren't a need-to-go-shopping wasteland; they're chock-full of the organic stuff of which we're made. The cosmos has gone green. Look into the darkness of a star-filled night sky, and some of the light you see has been tinged by passing through stellar clouds of sugars, fats—even alcohol. Reading *alcohol* might make you think of a shot of whiskey, a glass of wine, or a bottle of beer. What astrobiologists are seeing would keep all Earthlings tanked from here to eternity: clouds of diffuse cosmic booze millions of light-years across.

The mix of molecules discovered is a combination of, by Earth standards, the familiar, the rare, the exotic, and the truly bizarre. Given the astronomical extremes of temperature, density, and energies around stars and in interstellar space, there's a true astrochemistry that's different than anything seen on the relatively quiescent Earth—and in some cases it is so extreme that no one has yet been able to replicate it in a terrestrial lab. Out amid the stars are tongue-twister molecules like cyano-octatetra-yne and the butadiynyl radical, molecules that form in space but not on Earth.

Thus, astrochemistry is definitely its own chemistry. Yet what astrochemists have discovered is that this broader cosmic chemistry is different in particulars but not in fundamental processes from terrestrial chemistry. While there's enormous cosmic molecular diversity, what astrochemists have unveiled over almost half a century of cosmic molecular mining is that at their core, these cosmic molecules are ones we associate with us, with chemistry on Earth.

In a terrestrial context, it's not exactly a chemical stew you'd want to play in. The most common familiar cosmic molecules could be described as a pot of poison with a dash of sugar and a pinch of salt. Though toxic individually, what makes them deadly also makes them potent in life—they're highly reactive; they want to bond and form more complex molecules. After molecular hydrogen and water, the most common cosmic molecule is carbon monoxide, on Earth, an odorless, colorless gas released in combustion—from tailpipes and furnaces—that will quickly kill you if not vented properly. There's abundant ammonia, well-known as an ingredient in toilet-bowl cleaners; formaldehyde, also known as embalming fluid; and more than a touch of acetone, a common solvent used in nail-polish remover. There's also plenty of hydrogen cyanide, the toxic gas used in the Nazi death-camp gas chambers. Not to mention the abundant methane, a carbon with four hydrogen atoms, which, on Earth, is the extremely flammable primary component of natural gas. Amid this hoary chemical cocktail is a more welcome mix of substances. It turns out that interstellar space is awash with a molecule made up of two parts carbon, six hydrogen atoms, and a single oxygen atom to top it off. Chemists call it ethyl alcohol. Most of us celebrate it as vodka. There are the salts sodium chloride, the stuff of salt shakers, and potassium chloride, a key component of many agricultural fertilizers. Also acetic acid, named for the Latin word for vinegar, *acetum*, and the main ingredient of vinegar. To balance out all this alcohol and salt, there's also lots of simple sugar in the form of glycolaldehyde, an eight-atom molecule that's a building block for more complex sugars such as glucose, one of the sugars that makes table sugar.

One of the gradually building realizations of this cosmic carbon-chemistry awareness is that this chemistry also produces a flurry of big, complex carbon molecules, ones that on Earth are associated with barbecued steak and diesel exhaust. During the past three decades, as astrochemists cataloged bigger and bigger carbon molecules using radio-wave vibrations, over in the infrared there was evidence that something else was present in the interstellar stew. That something else now goes by the name polycyclic aromatic hydrocarbons, or PAHs, which are special for three reasons: they're big; they're rings; and they're everywhere, infiltrating every nook and cranny of the cosmos like the fine soot they resemble.

Rather than a single molecule, PAHs refers to a broad class of these molecules. PAHs are the molecular carbon giants of the interstellar medium. Whereas most cosmic molecules have few carbon atoms, the average PAH has fifty; some have many more. This places PAHs at the molecular borderlands, claimed as huge molecules by some astrochemists and as small dust by others. It's an academic hinterland all the more contested because PAHs are identified by the equivalent of a smudged molecular fingerprint—a broad band of infrared signals that points to an overall generic structure rather than to specific individual molecular ones.

Yet what defines them isn't so much their size as their shape. PAHs are thought to form from groupings of the ring-shaped carbon molecule benzene. Benzene is dubbed an "aromatic" molecule because when these ring-shaped carbon molecules were first identified by sniffing chemists in the nineteenth century, it was by their mildly sweet smell, which they'd only learn later is carcinogenic. Each carbon atom in the benzene ring, as well as clasping its carbon neighbors, is also adorned with a hydrogen atom, making the molecule a hydrocarbon—a molecule made up entirely of carbon and hydrogen. On Earth, when we hear *hydrocarbon*, we think petroleum and natural gas. The cosmic benzene rings join to form snowflake-like "polycyclic" structures, the PAHs.

Just as they're churning out other molecules, stars are pumping out PAHs. These big, sweet molecules are the most abundant organic molecules in the universe, accounting for up to one-tenth of all carbon atoms out there. They're everywhere, in almost every astrophysical condition, from the hottest to the coldest: PAHs are seen around dying stars, in the diffuse interstellar medium between stars, and in the ultracold dense molecular clouds that give birth to new stars. Viewed in the infrared, galaxies spew out vast clouds of PAHs into the intergalactic medium.

What makes PAHs all the more special is that they're far from being only extraterrestrial. On Earth, PAHs are usually thought of as pollution, formed from the incomplete burning of fuels. We cook up and eat PAHs in the carbon-y surface of whatever we're grilling when we fire up the barbecue, and we breathe them in when we're sitting in traffic behind a poorly tuned truck belching clouds of sooty exhaust. But thinking of PAHs only as pollution obscures a deeper truth: these cyclic carbon molecules are essential to life. Hydrocarbons come in two basic structures: rings and chains. The long chain-shaped hydrocarbons, such as the eight-carbon octane, are a key part of gasoline. It's when a chain of carbon atoms circle in on themselves to form stable rings that much of the most interesting biology takes place. For example, chlorophyll, the green pigment that enables plants to convert sunlight into sugars, is a large polycyclic molecule that includes several additional nitrogen atoms. Astrochemists have seen nitrogen-containing PAHs being spewed from dying stars, and many think there could be other, more complex PAH structures awaiting discovery. Even more essential, three of the nucleobases that form the backbones of DNA and RNA—cytosine, thymine, and uracil—are benzene-like ringed molecules in which two nitrogen atoms have replaced two of the carbon atoms. Thus, PAHs are more than a vast organic carbon repository; they're abundant evidence that the fundamental structural chemistry that underpins biology—the carbon-ring structure—is intimately wed to the cosmos. As a result, many astrochemists think that PAHs are a cosmic chemical feedstock—that these large molecules are eventually broken down by ultraviolet radiation and heat to provide the raw materials for further cosmic chemistry.

In our molecular searching, we are still at sea, still learning to interpret cosmic molecular fingerprints. All indications are that we have seen only the tip of the iceberg in terms of the molecular and mineral makeup of the interstellar medium, of what's actually out there between the stars. The strongest evidence for this is the existence of the enigmatic diffuse interstellar bands, the DIBs—one of the longest-standing mysteries in astronomy. When astronomers take the spectra of starlight that has traveled through clouds of dust and gas in the interstellar medium, they see not just the well-defined absorption and emission lines of identified atoms and molecules but also a forest of unidentified broad-absorption patterns, the DIBs. The term *diffuse* refers to the spectroscopic fact that these unidentified regions are broad valleys rather than thin, distinct lines. Paul Merrill, the Mount Wilson astronomer who would later provide the keystone observation that stars make elements, discovered the first DIBs in 1934. They've been a vexing, unsolved problem ever since. From the initial discovery of these mysterious light fingerprints, astronomers have identified more than three hundred DIBs in visual light, as well as dozens in the infrared region of the spectrum. The closer they look, the more they find.

There have been dozens of theories trying to explain what is producing DIBs. They have ranged from dust to molecules and from gas to solid materials. What all theorists agree on is that these deep spectroscopic prints can't be pinned on a single molecular suspect. Today, most astrochemists and cosmic-dust researchers are putting their money and telescope time on the chance that DIBs are produced by a range of large, complex carbon-based molecules. Other astronomers think that these "mysterious" carbon molecules might not be that mysterious, or at least not foreign, but that they are created by chains of amino-acid-like molecules, or proteins. Whether bloodlike pigments, muscle-like proteins, or the stuff of carbon-fiber bicycles, the DIBs represent a terrain of deep valleys and mountains within which lie further secrets of the Stardust Revolution.

The first astrochemists were struck by the possible relationship between cosmic chemistry and life on Earth. Charles Townes's discovery of water and ammonia alongside methane and carbon dioxide in interstellar space recalled the four molecules of a by-then famous experiment on the possible molecular origins of life. A decade earlier, Stanley Miller at the University of Chicago had mixed just those four molecules together, had zapped them with electricity, and, to his amazement, had produced a sludge loaded with amino acids. Townes and his colleagues saw that what Miller had done in the lab could perhaps occur throughout the cosmos. "These observations have an interesting connection with biology," the astronomers wrote in 1971, reflecting on the emerging view of a molecular cosmos.

But as much as this molecular cosmos resembled our molecular home, it was still *out there*—similar but removed by light-years from the messy business of life on Earth. Yet, while some astronomers were using radio telescopes to reveal a molecular cosmos, others had found that they didn't need telescopes at all. The next step in the Stardust Revolution would join *out there* and *right here* by using a whole new approach to stardust—not looking at it through a telescope but holding it in our hands.

PART 3

THE LIVING COSMOS

CHAPTER 7
CATCHING STARDUST

Would you rather look at stars through a telescope or hold stardust in your hands?

—Richard Herd, curator of the Canadian Meteorite Collection, 2011

THE SPACE-ROCK EDUCATION OF SCOTT SANDFORD

In astrobiologist Scott Sandford's cluttered office at NASA's Ames Research Center, twenty-five miles south of San Francisco, the walls are festooned with pictures of his scientific adventures. Some show him bundled up in a heavy parka during one of his Antarctic expeditions in search of meteorites. In his three-decade career, Sandford has stooped down to pick up pieces of the Moon—he found one of the largest lunar meteorites ever, MacAlpine Hills 88105—and was part of the team that from Antarctic ice retrieved a dark meteorite, Allan Hills 84001, identified as a long-traveling piece of Mars that some would later claim contained ancient fossilized Martian microbes. Yet for all this meteoritic success, Sandford is not in the picture of which he's proudest. He couldn't be. It's a simple black-and-white image of what look like tiny grains of sand—which is what they are, except that they were formed in our Solar System's nascent protostellar cloud more than 4.5 billion years ago, long before the existence of Cancun, Cape Cod, or any other terrestrial beach.

The grains were collected by NASA's Stardust mission, launched in 1999, for which Sandford was a leading member of the scientific team. Stardust was the beginning of a dramatically new kind of Solar System prospecting: collecting samples from their source. Rather than wait for asteroidal samples of the Solar System to arrive on Earth as meteorites, Stardust was sent out to sample a comet in its natural habitat—the first space probe to do so. By 2004, Stardust had caught up with comet Wild 2 at about Mars's distance from the Sun, where the probe sped to within one hundred and fifty miles of the two-and-a-half-mile-wide nucleus of the comet—close enough to pass through the comet's coma, the fuzzy halo of dust and ice ejected, geyser-like, from the comet's icy core as it approached the Sun. There, Stardust's aerogel collector—made of a material as light as air yet rigid and strong enough to stop microscopic comet shrapnel—swept up some of this microscopic cometary debris. In 2006, after an almost three-billion-mile round trip, the little explorer dropped off its extraterrestrial cargo in the Utah desert in a special return capsule.

Sandford was among the ecstatic scientists waiting there. After months of microscopic analysis, the Stardust team extracted thousands of tiny cometary particles, all in all less than a milligram of cometary material, weighing less than a grain of table salt. That might seem like a poor return for more than a decade of planning and a $200 million mission to bring back the first samples from beyond the Moon. But Sandford and the other Stardust scientists knew that size didn't really matter. These tiny grains held secrets to a much, much larger story, one that encompassed the entire origin of our Solar System, us included, in the stars.

What's all the more remarkable is that Sandford didn't set out on a career to study the stars. As a child, he'd been intrigued by astronomy, owned two telescopes, and once did daily drawings of sunspots for a month to see the Sun's rotation for himself. But growing up in Los Alamos, New Mexico, where his father was an engineer at the nuclear lab, Sandford was particularly captivated by the Earth's rocky bones, evident everywhere in the desert and mountains surrounding his hometown. He dreamed of being a geophysicist and did his undergraduate work at the New Mexico Institute of Mining and

Technology, spending summers working for a seismologist tracking earthquakes. But when Sandford applied to graduate school, his path turned skyward. He landed a fellowship at the McDonnell Center for the Space Sciences at the Washington University in St. Louis, Missouri, to study not terrestrial rocks but rocks from space.

Sandford began his research career just as the US space agency launched an interesting new program to collect dust from high in the Earth's atmosphere. The effort was spearheaded by Washington University geologist Donald Brownlee's radical contention that some of that high atmospheric dust wasn't Earthly effluent that had risen up but cosmic dust that was raining down. To test Brownlee's alien-dust idea, NASA agreed to bolt dust collectors onto the wings of some of its high-altitude U2 research planes—aircraft better known for their work as Cold War–era spy planes. "At that time, they had only been studying these grains returned from the U2s for a couple of years," Sandford tells me in his office at NASA-Ames. "When I joined the group, there was still no real definitive proof these things were even extraterrestrial, although we thought that many of them might be. They could have been smokestack effluent or volcanic dust that got up into the stratosphere." Working with colleagues, it was Sandford's job to prove the case.

To do this, he spent months of his graduate-school years in a lab clean room picking out bits of dust from the collectors. "St. Louis is often hot and humid," he recalls, "so frequently I'd go to the university with a bathing suit under my clothes. Then I'd strip down into the bathing suit and put on the bunny suit [the classic sterile clean-room coveralls] just so I wouldn't die of heat stroke." For hours on end, peering through a microscope, he used a micromanipulator to pick tiny bits of dust, whose size on average was ten to twenty microns (millionths of a meter) across—about the size of one of your red blood cells—out of the silicon oil in which they were trapped on the collectors. After he'd amassed this minute geological sample, Sandford used an electron-scanning microscope to take the sample's x-ray spectrum and to thus identify its elemental composition. For his PhD thesis in 1985, Sandford spent months comparing the infrared spectra, or light fingerprint, of the dust grains with the infrared spectra of comets and asteroids observed by astronomers. They were a near-perfect match—some with comets, others with asteroids. Brownlee was right; the dust grains Sandford was holding in his micromanipulator weren't of this Earth. They were bits of comets, asteroids, and maybe even the primordial stuff out of which the Solar System coalesced. At the time, it was a far-out idea.

"I remember the very first talk I gave at a science conference about this was attended by, I think, two people, because I was at a session with all the kooks who thought tektites [the glassy, rocky debris from huge meteorite strikes] were from the moon, which is wrong, and that the K/T boundary [the geological debris line associated with the end of the dinosaurs] marked an impact from an asteroid, which is probably right. But at the same time, none of those things were proven. So we were lumped in with the rest of the loony session, in a sense."

Today this daily cosmic dusting isn't considered crazy but is common sense. Most meteoritic material falls to the Earth not with a fireball's bang but as an invisible daily powdering. Every year, about forty thousand tons of these cosmic dust particles, averaging around twice the width of a hair, make landfall or splash imperceptibly into the sea. This cosmic dust comes primarily from two sources: meteoroids (the name given a meteorite while it is still in space) and interplanetary dust particles. If a meteoroid is smaller than your fist, the violence of its six-thousand-mile-an-hour entry into the Earth's upper atmosphere shreds it into a stream of cosmic dust more than thirty miles above the ground, and the dust gradually drifts down to the Earth's surface. Interplanetary dust is a combination of dust produced by pulverizing asteroid collisions in the asteroid belt between Mars and Jupiter and the minute debris of comets that evaporate or break apart as their orbit takes them close to the Sun, such as that collected by the Stardust mission. The Earth acts as a gravitational dust buster, sucking in these particles as it orbits the Sun. The result is that every day, on any sidewalk-square-sized area of the Earth's surface, whether in Manhattan or Mongolia, a tiny piece of outer space touches down.

Sandford was one of the first people to see this firsthand. Through his extraterrestrial dust research, the young man who'd wanted to study the dynamics of the Earth's rocks was initiated into a new field of the Stardust Revolution: astrogeology, or lithic astronomy. Astrogeologists watch stars not through telescopes but through microscopes. A century and a half earlier, the French philosopher of science Auguste Comte had dismissed astronomy's future as a science because, he argued, you'd never be able to actually hold a piece of star in your hand and do experiments on it to determine its fundamental nature. But Scott Sandford has done this, and now geology is joined directly with the stars. "We normally think of everything around us being ultimately four and a half billion years old because that's when it all came together," he says.

> But of course every atom itself was outside the Solar System and came together when it formed. Even more intriguing is that we didn't just get a delivery from the interstellar medium in the form of atoms. We got molecules and grains, and some of them survived relatively unscathed. As individual entities, they are older than the Solar System. If you want to understand how our Solar System formed, there is a limited amount you can learn from planetary materials like Earth rocks because they have all been completely reprocessed. Because of planetary geological processes, there are no rocks on the Earth that witnessed the birth of the Earth. If you're looking for samples of primitive primordial material, the starting stuff that all of us began with, you need to look to meteorites and comets.

The first great Stardust Revolution story that meteorites would tell was of the birth of the Earth.

THE BIRTH OF THE EARTH

Taking a walk with a toddler, you get a lot of time to think about how amazing rocks are. When my son was a two-year-old, we couldn't walk more than ten steps before he was bending down, picking up a nondescript bit of crushed gravel, and holding it up to me with a look of deep joy. "Rock!" he'd exclaim. We all had this sense of wonder about the rocks around us, though most of us have lost it. But if there was one rock in the twentieth century that really made planetary geologists and astronomers excited—that had them holding it up to the rest of the world and shouting, "Look!"—it was the Allende meteorite, a rock that opened a new chapter in the Stardust Revolution.

Meteorites are often named after the closest post office by which they're found. With the Allende meteorite, coming up with a name was simple. Around one in the morning on February 8, 1969, in Pueblito de Allende, in the northern Mexican state of Chihuahua, one of the chunks from a massive shattered space rock missed crashing into the local post office by just thirty feet. Allende was no make-a-wish shooting star. It collided with the Earth with the force and momentary terror evoked in a Hollywood cosmic-collision apocalyptic blockbuster. The sedan-sized meteoroid entered the Earth's atmosphere at supersonic speed, creating a series of thunder-like sonic booms that awoke sleepers from southern Arizona into northern Mexico. Its blue-white fireball "was so bright we had to shield our eyes," said Guillermo Asunsolo, the editor of a Chihuahua newspaper. "The light was so brilliant we could see an ant walking on the floor." The terror evoked by this midnight intruder sent many rural Mexicans running for the local Catholic churches. "The people, especially the people in the small villages are very alarmed," Asunsolo reported. "They say that this is an announcement that the world will soon end."

The Allende meteorite, however, turned out to be an amazing heaven-sent messenger revealing details not about the Earth's end but about its beginnings. For the first time, scientists would realize that they were holding stardust in their hands and gaining an understanding from this stardust of the forces that shaped the Earth's birth—and of a time when the Earth was part of a cosmic cloud of dust, gas, molecules, and ices holding all the potential of its next billions of years, yet it was not even a blue glimmer on the cosmic scene.

Allende had arrived with perfect timing. In preparation for the Moon rocks to be returned by the Apollo 11 mission just months later, NASA had prepared a national network of laboratorics with the latest geochemical-analysis tools to study the first lunar samples. Before Neil Armstrong and his crew returned the first lunar rocks, the Allende meteorite delivered a free payload of cosmic material that would turn out to be as scientifically important as the Moon rocks. Within hours of Allende's impact, meteorite scientists were rushing to the site to gather possible fragments. The first to arrive at the impact site was Bert King, a NASA scientist who'd been busily preparing the Lunar Receiving Laboratory in Houston, and who would soon become the first lunar sample curator. King could be away from work only for little more than a day. He didn't sleep for thirty hours, consumed as he was in a meteorite-collecting spree that had him buying pieces of the Allende meteorite from locals with the help of a Mexican policeman as interpreter and negotiator.

What was most important, though, was his reconnaissance for those who'd follow. Colleagues at the Smithsonian Institution in Washington managed to track him down by phone, and he told them of the wealth of meteoritic material that lay strewn across the remote Chihuahuan semidesert. During the next several weeks, Smithsonian scientists enlisted a local search party, including many schoolchildren, to collect meteorite fragments. Together they collected more than two thousand pounds of the Allende meteorite, about 2,100 pieces, ranging in weight from as little as several ounces up to a meteoritic whopper of 242 pounds. The pieces were strewn over a debris field of nearly two hundred square miles, the result of a literal meteorite shower when it had exploded.

The large number of pieces was crucial to what would follow. With an abundance of material, the Smithsonian readily distributed bits of the meteorite to interested researchers. They were able to destroy pieces in their analysis, if necessary, without worrying about running out of priceless research material, as is sometimes the case with rare samples. As the results of the Allende meteorite analysis arrived, they showed that this rock from space packed more than a shock for just the Mexican villagers who'd experienced its spectacular arrival. It has continued to surprise scientists for half a century.

For Bert King, the greatest surprise was the meteorite's type. To put his excitement in perspective, it's important to know that there are three categories of meteorites, each defined by composition and asteroidal origins: iron, stony iron, and stony meteorites. And just by its composition, each of these types of meteorites tells a story. "When geologists look at a rock they don't just say 'That's a sandstone' or 'That's a granite,'" meteorite scientist Richard Herd tells me when I visit Canada's National Meteorite Collection in Ottawa.

> What happens is that immediately the wheels start turning, and they try and figure out how the rock was made, under what conditions. Is this a fluvial sandstone formed in a river; is it a beach deposit; is it a desert sandstone? Because a sandstone is literally a rock made out of sand. So the kind of material that a rock is made of and the way that material is put together—whether it's igneous, metamorphic, or sedimentary—tells you something about the environment of formation. That way we can read the history of the Earth. Meteorites are rocks that reflect the processes that they've been through, just like the Earth rocks. They have a history and provenance that can be deciphered. These geological processes tell you a story.

Iron meteorites get their name from what they are made of, which is readily apparent when you hold one in your hands and feel a quick sense of preparing to hold something heavier than it looks: solid, heavy, slag-like chunks of iron and various amounts of nickel and other siderophile (iron-loving) elements. As such, they'd be better named steel. Cut one crosswise, and you end up with a slab that looks as if it could have been rolled straight out of a steel mill, but that lump of iron is a sampling of a shattered asteroid's core. The asteroid was large enough that, as occurred in the Earth, its interior melted (in the asteroid's case, through radioactive heating early in its life), and its heaviest elements, iron and nickel, formed a metallic core that slowly solidified.

The second class of meteorites are the stony irons. As their name implies, these meteorites combine the qualities of iron meteorites with terrestrial rocklike characteristics. The stony parts of these meteorites are the remains of the crustal, rocky portions of asteroids. Asteroids larger than about fifty kilometers in diameter—called planetesimals, denoting their mini-planet-like natures—developed into differentiated, layered mini-worlds with molten cores and solid crusts, just like Earth. During its hot, formative early years, an asteroid's geology had many similarities to terrestrial geology. The stony, crustal meteorite bits are cosmic cousins to both terrestrial basaltic rocks (those formed by lava) and granites (those igneous rocks formed when molten rock slowly cools below the surface). The stony irons contain the complex story of when asteroids collided, producing impact-generated mixes of interior and crustal rocks.

The stony meteorites are the third group of these rocks from space. It's a subclass of stony meteorite, the chondrites, that are the core and the future of the Stardust Revolution. Chondrites are the sedimentary rocks of the solar system. They're agglomerations of cosmic sediments, the original particles present in the solar nebula and protoplanetary disk. Remove the Sun's hydrogen and helium, and the mix of elements you'd be left with would resemble those of a chondrite. Chondrites are characterized by containing chondrules, a word derived from the ancient Greek word for grain. Chondrules are millimeter-to-centimeter-sized glassy, metallic beads that formed from rapidly cooling metallic gas, free-floating molten droplets, in the emergent blast furnace of dust and gas surrounding the early Sun. They are the Solar System's first igneous rocks.

What gets stardust scientists really excited are the rare carbonaceous chondrites, meteorites that are rich not only in the elements that make the Earth's core and crust, but also in the element that's central to life: carbon. Some carbonaceous chondrites are so rich in carbon-based molecules that they resemble charcoal briquettes or even peat. What makes the carbonaceous chondrites even more special is not how they've changed but how they've stayed the same. They are the least-altered meteorites, the ones that preserve the nature of the early protoplanetary disk from which all that is the Earth formed. The asteroids, of which meteorites are a chip off the larger block, formed within the first ten million years of the Solar System's birth. That's about fifty million years before the Earth-Moon is thought to have formed from the massive collision and fusing of a couple of other planetary embryos. Each carbonaceous chondrite is a tiny sampling of the natal protoplanetary disk material from which life emerged. As such, each carbonaceous chondrite is greeted with profound interest by stardust scientists, who seek to tease every detail about the origins of life from each new immigrant from outer space.

Looking at the thirty-pound chunk of the Allende meteorite on a newspaper editor's desk in the village of Hidalgo del Parral, Bert King knew he was looking at just such a carbonaceous chondrite. The fresh chuck of Allende that King saw was so rich in carbon he could smell its oily, organic nature. Today these meteorites make up only about 4 percent of all collected meteorites, and Allende still holds the record as the largest carbonaceous-chondrite meteorite shower ever.

The biggest surprises were waiting in the labs of King and other geologists. From broken pieces of the meteorite, King could clearly see that it contained a mix of distinctive white circular and fuzzy-edged blobs amid the meteorite's darker matrix, separate from and less abundant than the chondrules. These whitish inclusions, most of them much less than a penny's diameter in size, turned out to be mother lodes of calcium and aluminum; thus they were dubbed calcium-aluminum–rich inclusions. What makes these inclusions truly special is their age. Using uranium-lead dating, geologists determined that the calcium-aluminum–rich inclusions were older than the Earth.

Here was a seemingly impossible idea. How could something be older than the Earth? The Allende meteorite is so special because it's an amalgam of materials that coalesced thirty million years *before* the Earth formed, when the nascent Sun was surrounded by the still-developing protoplanetary disk. The calcium-aluminum–rich inclusions formed even before that. They were the first solid objects to emerge in the hottest areas of the solar nebula, marking the beginnings of the solid phase of our Solar System. Whereas most materials were still molten or gaseous, the calcium-aluminum–rich inclusions had a high enough melting point, almost 2,600°F, that they were the first objects to condense out of the cooling solar nebula. In some of the most detailed uranium-lead dating ever, scientists have pinpointed the age of some calcium-aluminum–rich inclusions to 4.567 billion years old, give or take half a million years. The Allende meteorite contains some of the first solid materials to form in our Solar System; tiny time capsules from the very beginning of what we call home.

The revelation of Allende's age, and particularly that of the calcium-aluminum–rich inclusions, set astronomers and geologists on a quest to see what else they could learn about the Earth's prenatal stage. This was a truly remarkable thought—it was possible to perform hands-on cosmic archaeology and gain information about the Earth's origins not from terrestrial rocks but from rocks originating elsewhere in the Solar System. The Allende meteorite isn't alien material but rather is sibling material from the same stellar nursery. As an older sibling, it can tell us stories about the world before our arrival. For 4.567 billion years, the Earth and the Allende meteorite's asteroidal parent performed a mutual orbital dance around our Sun, until at some point the asteroid smashed into another body, ejecting a piece of asteroidal shrapnel that on a winter's night in 1969 became a shooting star. With its fiery impact, Allende brought to light the story of the Earth's birth.

With the insights of the astronomer's periodic table—the ways that particular stars forge particular elements—astrogeologists turned to analyzing Allende's chemical makeup for clues to its and, by extension, the Solar System's origins. The Allende meteorite received the most intensive chemical fingerprinting of any rock ever. The most intriguing thing this analysis revealed was that those whitish calcium-aluminum–rich inclusions had oddly abundant levels of the isotope magnesium-26. This particular form of magnesium is a decay product of the radioactive isotope aluminum-26. Aluminium-26 in turn is formed in the blink of an eye during a supernova when enormous temperatures ignite the fusing of neon, producing a supernova spray of aluminum-26. But aluminum-26 has a half-life of only about seven hundred thousand years. This means that a rich burst of aluminum-26 seeded the solar nebula just before the calcium-aluminum–rich inclusions formed.

In effect, magnesium-26, the fossil remains of aluminum-26, appears to be the smoking gun that proves that the shock from a nearby supernova might have triggered the gravitational collapse of the solar nebula that formed our Solar System. Although today the Sun and its stellar siblings are all far from their natal home, it's thought that the Sun, like most stars, formed as part of a densely packed stellar cluster—a family of hundreds of stars formed from the same dense molecular cloud, such as we see occurring today in the Orion Nebula. Amid this stellar family was at least one, and probably several, short-lived giant star, which, perhaps before the Sun had shed its first light, detonated, spewing energy and elements, including aluminium-26, for hundreds of light-years; the shock triggered the collapse of the dense molecular core from which emerged our Solar System. This evidence of supernova midwifery has stood the test of time and further scientific scrutiny. It appears that the death of one star sparked the birth of another, a phenomenon astronomers see played out in stellar nurseries across the Milky Way.

In less than a decade of its arrival, the Allende meteorite had pushed back the veil of time to reveal turbulent lands and awesome events that were the equivalent of the moment of our Solar System's conception. However, Allende held deeper secrets, ones that would require even greater patience and persistence to tease out but that would take us back even farther in time, to reveal the direct stellar roots of our family tree.

THE MEN WHO FIRST HELD STARDUST

On his office desk in the University of Chicago's Enrico Fermi Institute, Edward Anders kept a set of red-capped, wooden Russian dolls; hollow figures that nestle inside each other, each one smaller than the next; the final, tiniest one being solid. Years later, Anders would reflect that the symbolic answer to his quest had been right in front of him as he'd made calculations and discussed strategy with his research team. But that observation was made in hindsight. At the time, Anders, an avowed master of meteorite studies, was stumped. There was something chemically strange about the Allende meteorite, but in the mid-1980s, after more than a decade of searching, Anders's team couldn't open the final layer of the Russian doll and see what it was.

Anders was used to tough research questions and had made a name for himself as an impassioned intellectual fighter who'd spar with all comers. In 1964, he published a landmark paper that laid out the modern understanding of meteorites' origins: they're chunks of wayward asteroids from the asteroid belt between Mars and Jupiter. The problem was that Anders's assertion was a full-on challenge to the view of leading planetary scientist and Nobel laureate Harold Urey, his former mentor, whose post at the University of Chicago Anders had filled in 1955. Urey argued vehemently that meteorites were blasted-off bits of the Moon rather than original Solar System material.

Anders's first paper proposing the asteroidal origin of meteorites was rejected by the leading astrophysical journal. He received a letter from a respected meteorite researcher telling him that his asteroidal model was ridiculous. In his 1964 paper "Origin, Age, and Composition of Meteorites," Anders pored over, correlated, and stitched together thousands of pieces of varied meteorite evidence available at that time from the fields of chemistry, physics, mineralogy, and astronomy. His wife complained that he worked every day in December 1963 while he was on sabbatical at the University of Bern in Switzerland. Anders sealed the case, claiming that all the evidence pointed to the fact that the parent bodies of meteorites weren't other planets or the Moon but rather were asteroids. He was right. Martian and lunar meteorites may get lots of press, but these rocks actually account for only about 0.5 percent of all meteorites.

While Edward Anders was describing the origin of meteorites, in the Soviet Union, the planetary geologist Viktor Safronov was using meteorites to piece together the seemingly unreachable origins of the Solar System. Safronov had taken the intellectual baton from his mentor, the theoretical geologist and mathematician Otto Schmidt. Purged from the Soviet scientific establishment by Joseph Stalin in 1942 because his father was German, the unemployed Schmidt turned to a detailed imagining of the cosmic origins of the Solar System. Although the United States won the race to set foot on the Moon, it was the Soviet Union, using Schmidt's pioneering work, that won the race to figure out how the Earth, the Moon, and rest of the Solar System formed. In 1969, Safronov published his book *Evolution of the Protoplanetary Cloud and Formation of the Earth and Planets*, the name of which in the original Russian sounds quite different, but the concept of which is identical to the 1972 English translation. In this book, Safronov synthesized more than a century of reflection into one critical insight: stars don't form alone. Stars are born, said Safronov, surrounded by a vast swarm of gas and dust that over time gravitationally settles down into a circumstellar disk, a thin band of material orbiting its star like the swirling, uplifted skirt of a spinning dancer. Planets, and everything on them, are made from these star-making leftovers.

Since Safronov's seminal treatise, stardust scientists have filled in the details of this grand vision. At the core of their findings is a profound conclusion: the Earth emerged from stardust. Every mineral and molecule—from green olivine to formaldehyde—that coalesced to make the Earth and the other bodies of the Solar System was forged by existing stars and processed in the chemical mash-up that is the interstellar medium. Safronov provided a framework for extending terrestrial geology into space, for understanding our Solar System as if it were a single rock in whose multitudinous layers, minerals, and elements could be read the story of its origins.

It was sentences of this ancient story that Anders was trying to read in bits of the Allende meteorite, but the message contained in the space rock's isotopic signature had Anders stumped. Isotopes of an element are like identical twins—exactly identical to most observers, but to someone who knows them well, they are different in important ways. All atoms of an element have the same number of protons and electrons—and, thus, the same chemical properties. But some atoms of an element are a little heavier than others, containing one or more additional neutrons. Chemists use a mass spectrometer to identify and sort isotopes of the same element based on differences in mass. The relative distributions of various elemental isotopes in a mineral or molecule—its isotopic fingerprint—can provide a powerful clue as to its origins. In the Stardust Revolution, this isotopic fingerprinting has become transformative in creating a detailed family tree of the ancestry of cosmic minerals and molecules—the field of stable isotope cosmochemistry. By the early 1970s, Anders had the world-leading cosmochemistry lab, attracting the most talented and ambitious graduate students and postgraduate researchers to his team, with Anders acting as the intellectual quarterback.

With Allende, Anders was trying to crack the ultimate forensic cold case—what he called the meteorite's mysterious isotopically anomalous xenon component. This isotopic mystery involved one of the least common elements on Earth: xenon, from the Greek *xenos*, "strange." In 1964, meteorite researchers discovered that Allende contained anomalous isotope levels of xenon. Of xenon's nine stable isotopes, the meteorite contained at least double the amount of both the lightest and heaviest isotopes found in terrestrial samples. Xenon is not just rare; it's also a "noble gas," essentially a loner element, along with krypton, neon, argon, and helium, that has all the electrons it needs, and so it doesn't chemically react, or bond, with other atoms. It's also a gas at room temperature, and since the meteorite had been stored at this temperature, the fact that the xenon hadn't off-gassed meant that it must be physically trapped in the meteorite. Allende's isotopic mix was tantalizingly distinct from that seen in other meteorites, and this pointed to some astrophysical process that had shaped the meteorite, the memory of which was recorded in the xenon. There was a stranger in the meteorite. The question was, where was it hiding?

Figure 7.1. On his way to helping discover true stardust, Holocaust survivor Edward Anders holds the Hamlet Indiana meteorite in 1959 at the University of Chicago. *Photo reproduced with permission from the Special Collections Research Center, University of Chicago Library.*

That Anders was around to contemplate the origin of the xenon was itself an anomaly. Anders was born Edward Alperovičs on June 21, 1926, the summer solstice, in the town of Liepāja, Latvia, where his grandfather Israel was head rabbi. When Nazi forces attacked the Soviet Union and entered Latvia in June 1941, there were approximately ninety thousand Jews living there. Only 2 percent of them survived the Nazi occupation. What occurred in Liepāja was similar to what happened in other Latvian towns. In late December 1941, an SS *Einsatzgruppe*, or mobile death squad, arrived in Liepāja. Jews were rounded up and marched to a former Latvian firing range on rolling sand dunes beside the sea. The first to arrive were forced to dig a mass grave. Then, over the course of three days, almost every Jew in Liepāja was murdered. The scar of the shoreline mass grave is still visible from the air. Twenty-four members of Anders's extended family, including his father Adolph and brother Georg, were shot on those dunes. Of the Anders family, only Edward and his mother, Erika, survived—the result of a ruse whereby the family had agreed that Erika would claim she was an Aryan foundling raised by kindly Jews, and thus her children were only part Jewish. Surviving the war, Anders arrived in the United States in June 1949, working as a waiter in a downscale Jewish hotel in New York's Catskill Mountains until his acceptance at Columbia University. He quickly excelled there and became intrigued with meteorites, to whose study he devoted the rest of his research career.

In 1972, Anders was joined by postdoctoral researcher Roy Lewis, who'd trained with the world's leading mass-spectroscopy expert and now brought his isotope-weighing skills to Anders's lab, taking over much of the lab's hands-on research, which by 1975 Anders had stopped, in favor of performing the reading, coaching, and paper writing of a senior scientist. In search of the location of Allende's xenon, Anders, Lewis, and several graduate students developed a procedure in which a piece of the meteorite was subjected to a series of progressively harsher acid and bleach baths that gradually dissolved most of the meteorite—particularly its rocky matrix and the chondrules. After each bath, the researchers used mass spectrometry to analyze the remaining acid-resistant solid residue for xenon. If it was present, they'd continue on a laborious, trial-and-error journey and choose a new corrosive solution, such as concentrated ammonia, chloric acid, or hydrogen fluoride, any one of which was strong enough to further eat away at the meteorite's remains. Even after a decade of on-and-off work, however, they couldn't identify exactly where the xenon was trapped.

The breakthrough came serendipitously in the spring of 1986, when Anders, Lewis, PhD student Tang Ming, and postdoctoral researcher John Wacker found what they were looking for. They'd reduced a sample of the Allende meteorite down to a mere 0.5 percent and were left with a black, tarry residue that still contained the xenon. In an attempt to further reduce the sample, Wacker set up an overnight treatment in which he mixed some residue in a powerful acid, put the mixture on a hot plate at almost 300°F, and went home. But the thermostat on the hot plate got stuck, causing the sample to overheat, and when Wacker returned in the morning, his sample had turned white. At first, Wacker thought he'd ruined the experiment, since all known forms of carbon, except one, are black. But Lewis decided to run with what they had and to try to identify the white powder using x-ray diffraction.

Small samples of a crystal can often be identified by x-ray diffraction; each mineral bends or diffracts light in a characteristic way to create a kind of diffraction shadow—the rough equivalent of how a human skeletal pattern appears on an x-ray. When Lewis subjected his minuscule space-rock samples to this x-ray examination, the diffraction shadow that appeared was of an uncannily well-known terrestrial material, one he'd never imagined finding in a meteorite: diamond. There were trillions of them—what we now call nano-diamonds—each one too small to be measured in carats; rather, they could be measured only in terms of carbon atoms, averaging about two thousand atoms. "They would barely make engagement rings for bacteria," Anders quipped.

No matter how small they were, they were still *diamonds*—something that, Anders and his coauthors wrote, "no one had seriously considered a constituent of the cosmos outside the Solar System." Geologists thought that diamonds could be created only by the high temperatures and pressures inside planets. In fact, astronomers using infrared telescopes had first detected what looked like the infrared light fingerprints of interstellar diamonds as early as the late 1960s. They had discounted the detections as impossible. Clearly the diamond-formation theory needed tweaking: Where could this collection of cosmic jewels infused with anomalous amounts of xenon in a meteorite come from?

If the diamonds were a surprise, the answer to their origins was even more bewildering. So much so that, even years later, scientists reflecting on the discovery noted that it could reasonably seem to strain credulity. Anders and his colleagues saw only one feasible source for these cosmic diamonds: pure stardust. At the time, astrophysicists thought it was impossible that raw stardust could survive from one generation of stars to the next. Common sense indicated that this primordial dust would inevitably be mixed and lost in the maelstrom of collisions, heating, nuclear reactions, and intense radiation of star birth. But the nano-diamonds in their samples weren't reprocessed protoplanetary debris that had been morphed, melted, or otherwise changed. They were the pure, original output of stars. And not just any stars. Now the xenon made sense. The diamonds had condensed out in the carbon-rich atmosphere of a red giant, a medium-sized star in its bloated death throes. Later, when the star's remnant white dwarf core exploded as a Type 1a supernova, the stellar diamonds were infused with a spray of xenon atoms, which they'd carried ever since.

Anders and his colleagues hadn't just dissolved a meteorite to find stardust; they'd dissolved time. The sequential acid baths had mimicked our Solar System's accretion process in reverse. Instead of building up from stardust, the process used by Anders's team had worked back to it. They'd taken a meteorite, itself a chunk of asteroid, and winnowed it down to its most fundamental essence: stardust. The billionth-of-a-meter-sized diamonds were the oldest minerals anyone had ever discovered, let alone ever imagined. They weren't just stardust. They were stardust from stars that lived and died and whose dust was dispersed through the galaxy long before our Solar System was born. This was pre-solar stardust. In their lab glassware were the remains of long-extinguished stars that had burned bright somewhere across the Milky Way. Their light was gone, but their diamond dust remained. As a scientist, Anders was renowned for an elephantine memory, but this was a whole new kind of remembering. The teenager who'd survived some of the bleakest, most murderous days of the Holocaust had helped find slivers of preserved ancient starlight amid cosmic darkness.

"As a young graduate student I was bowled over by the sight of meteorites that came from one or two astronomical units [the Earth–Sun distance] beyond Earth," Anders later recalled. "Now, as an old man, I am awed by the sight of stardust from a few hundred light-years beyond the Earth; ten million times further. Such is the magic of meteoritics."

STARDUST MEMORIES

Anders and his colleagues had forged a new field of the Stardust Revolution: hands-on astrophysics. Just as the emerging nuclear physics of the 1940s and '50s intersected with the understanding of nuclear processes in stars, stardust geology is merging with astronomy to provide what geologists call the ground-truth of stellar processes: you don't just look or talk about it; you can actually "hold" and examine up close a piece of stardust. Meteorites, Anders said, were the poor man's space probe, able to deliver samples directly from a star.

"Until recently people could study stars only by remote means," he told a reporter just after the discovery. "Astronomers who have been studying stardust for years with their telescopes could make only some general statements about the chemistry and the grain size and so on, considering the nearest sizable quantity of stardust is a hundred light years away. No one had looked at any samples. But we have made it possible to study stardust with all the techniques of modern science."

Stardust geology—a subdiscipline of astrogeology—is the study of rocks at the smallest, oldest, and most primordial level possible; it's the study of grains belched directly from a star. Unlike metaphysical stardust, these stellar crystals aren't all pretty and glistening. They look more like bits of beach sand that you'd shake out of your shoe after an oceanside walk: angular, pocked cubes of silicon carbide or crusty balls of graphite. What's remarkable is that the same geological principles apply in understanding stardust as they do in understanding a chunk of coal from Appalachia or the freshest piece of volcanic basalt. Each grain's structure and composition, its shape and what it's made of, are the results of the formative processes—the heating and cooling, pressures, and elemental merging—that made it. It's just that, rather than telling the terrestrial story of fine sediments accumulating over some tropical reef or of the gradual heating and crushing of limestone deep in the Earth to form marble, stardust reveals the story of the particular nuclear reactions inside a star and then the turbulent processes as these elements spew forth, bond, merge, and gradually cool into the first rocks: stardust. And just as geology reveals the history of the Earth, this lithic astronomy reveals the history of the cosmos—but before the Earth existed.

Pre-solar grains are the cosmos' oldest artifacts. Each is between five and seven billion years old. It's difficult to pinpoint the ages more precisely, because each grain is so small one has yet to be radioactively dated. What does age mean for a rock? On one level, everything in the cosmos is the same age, all of it originating with the big bang. A discussion about the age of terrestrial rocks, stardust grains, or ancient Egyptian artifacts refers to how long something has existed in its current state. For example, your age isn't the age of your atoms but rather is the age of the time since they joined to form you. The same concept applies to stardust grains. Cosmic-dust scientists think that the vast majority of stardust is reprocessed numerous times before forming part of a solar system. For example, just as the Earth's continents go through a gradual cycle of movement and mixing—the plate tectonics that cause earthquakes, volcanoes, and tsunamis—stardust goes through an endless process of physical change. It's transformed by heat, cold, pressure, dramatic dust collisions and mixing, and repeated cycles of all these elements. It's obliterated—vaporized by supernova blast waves—then locked into planets for billions of years, only to be recycled into the interstellar medium with a star's death—either slow and billowing or with blink-of-an-eye explosive force. Although it certainly doesn't sound as impressive on an inspirational poster, we are made not so much directly from stardust as we are from heavily reprocessed interstellar dust. In contrast, pre-solar stardust grains are truly the unchanged remains of our ancestors, and, as such, they hold the deepest memories of galactic history and our cosmic origins.

Stardust geologists work on a previously inconceivable scale. For these stellar rock hounds, the emblematic tool of the trade isn't the geologist's rock hammer but rather is the nanoSIMS—nano Secondary Ion Mass Spectrometer. In the nanoSIMS process, a stardust grain is bombarded with a high-energy ion beam that gradually erodes the sample, as occurs in sandblasting, sending atoms hurtling off from the sample and into the mass spectrometer, a sophisticated scale that measures atomic masses by tallying the way they respond to electric and magnetic fields. The end result is that nanoSIMS produces a grain's detailed atomic makeup, its atomic fingerprint. What's critical are the isotopic details of this stardust fingerprint. They're distinct from the mixed material of our Solar System in almost every element. To a stardust scientist, isotopic anomalies, like Anders's xenon anomaly, are the hard evidence that prove that a bit of microscopic dust is indeed the stuff of a star.

While all this is awe-inspiring, what's even more remarkable is that with this nanoSIMS grain-by-grain isotopic fingerprinting, stardust scientists can trace a grain's parentage not just to the stars but to particular kinds of stars at particular stages of their lives. It's one grain, one star type. The isotope differences start with the parent star, and at each stage of its life, a star has a distinctive isotopic fingerprint. Depending on the temperature, pressure, and elemental mix in a star, different atomic nuclei are formed. A supernova produces a different mix of isotopes than does an old red giant. Eventually, these stellar isotopic characteristics are frozen in stardust, which in turn, billions of years later, holds its story accreted inside a meteorite. Differences in isotope ratios can range from a modest parts-per-thousand to in-your-face thousandfold differences. For example, the average ratio of oxygen-18 to oxygen-17 in the interstellar medium is different from that ratio as it exists in the Solar System, a fact that provides cosmochemists with a clear isotopic signature that reveals whether a mineral or a molecule was formed in the interstellar medium or after the formation of our solar nebula.

Figure 7.2. A pre-solar grain of silicon carbide extracted from the Murchison meteorite. About one-millionth of a meter in size, the grain's isotopic signature links it to the death of a red giant star. *Image by Scott Messenger, reprinted from* Geochimica et Cosmochimica Acta *67, no. 24, Thomas J. Bernatowicz et al., "Pristine Presolar Silicon Carbide," 2003, with permission from Elsevier.*

Most of the work on tying stardust back to parent stars has been done with the best-studied stardust type, silicon carbide, a crystalline combination of carbon and silicon. Silicon carbide is one tough material. It was the second stardust mineral that the University of Chicago researchers isolated and was able to survive numerous rounds of alternating acid-bleach corrosive attacks. On Earth, silicon carbide rarely forms naturally, but it is synthesized for its diamond-like hardness and durability; for example, as the grit on sandpaper and part of the bullet-stopping material in body armor. This ability to withstand forces that dissolve or abrade other minerals makes it a great cosmic storyteller. Stardust scientists have isotopically fingerprinted thousands of individual grains of silicon carbide from meteorites and, in the process, have linked these particles back to two types of dying stars: supernovas and red giants. Nine in ten silicon carbide stardust grains are thought to emanate from red giants. Similarly, some silicon carbide grains carry their supernova origins in their isotopes. Some of these grains have large excesses of calcium-44 relative to calcium-40. Once again, the reason is radioactive decay. Supernovas produce large amounts of the radioactive isotope titanium-44, which quickly decays to calcium-44. Only exploding stars forge abundant titanium-44, and thus the presence of its calcium decay product is the evidence that a grain of stardust formed in the cooling outflow of a cosmic-scale firecracker.

Today, stardust researchers have amassed a catalog of more than ten thousand pre-solar grains that have been isotopically fingerprinted back to their natal star type, including at least a dozen different pre-solar minerals. Along with nano-diamonds and silicon carbide, these pre-solar minerals include titanium carbide, globular balls of graphite and grains of various silicate minerals—otherwise known as sand, which are formed in the atmospheres of red giant stars and are the type of mineral from which most of the Earth's crust is made. From the diversity of stardust grains, Edward Anders once ballparked a guess that our Solar System formed from the stellar remains of at least one thousand dead ancestral stars. We see only

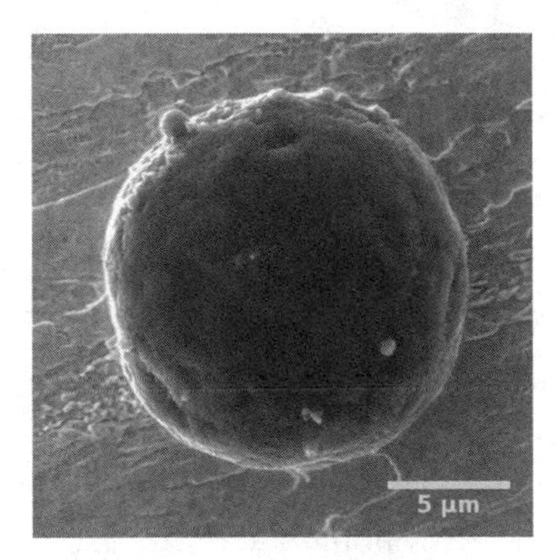

Figure 7.3. Older than the Sun, this grain of graphite—or pencil lead—isolated from the Orgueil carbonaceous chondrite is stardust formed from a long-ago supernova. *Image courtesy of Evan Groopman, Washington University, St. Louis, Missouri.*

part of the evidence in pure stardust. Most of the atoms from those stars have been thoroughly recycled, their pedigree lost in the general mix of our Solar System. The story of our origins is being pieced together stardust grain by stardust grain.

These were the kinds of pre-solar grains from which the Earth formed. As you read this, careers are being spent—days and weeks of computer modeling, months of space-satellite observations, painstaking hours with electron microscopes probing otherwise invisible bits of stardust—trying to piece together the exact steps by which stardust grew to become the Earth and other planets. What's clear is that, in rough outline, the process was a case of a great cosmic clumping. The story of our cosmic origins is one of coagulation, of the coming together of, for all intents and purposes, an infinity of otherwise negligible cosmic bits and pieces. It began when, inside a giant molecular cloud, some gravitational instability, a ripple in space-time, caused the cold dust and molecular gas—all the gas is molecular when it's at just a smidgen above absolute zero—to start to clump and spiral in on itself. It was the first contraction on the way to the birth of a solar system. Cosmic matter spread over a diameter of perhaps ten light-years—

a bubble sixty trillion miles across—collapsed down to a mere five billion miles across, the approximate distance from the Sun to the outermost reaches of our present-day Solar System. About 99 percent of this cloud was gas, most of it hydrogen, with a smaller mix of helium. The bulk went into forming the Sun, but there was also a solar nebula, a cloud of surrounding material of between 1 and 7 percent the mass of the Sun. Most of this material was also gas—but a tiny fraction, just 1 percent, was molecular and mineral dust. As this gas and dust swirled around the gestating Sun, it formed into a flattened frisbee-like disk around the star, what astronomers call an accretion disk, providing the first inkling of the Solar System to come. Through microscopic dust collisions, slightly larger particles took shape. It might have taken millions of years of orbiting for a stardust grain to grow to the size of a marble and for these marbles to merge into boulders, but time is all that was needed. As the grains grew, the process accelerated. More massive chunks of material were gravitationally drawn toward one another and toward the center of the accretion disk, creating a denser cluster of orbiting traffic. The collisions were complicated by the baby stars' growing-pain outbursts of radiation; caught up in currents of radiation and magnetic fluxes, grains were melted, fused, evaporated, sublimated, irradiated—repeatedly heated and cooled in a maelstrom of creation.

For all its cosmic proportions, the birth of the Solar System recognizable to us today occurred within the blink of a geological eye. Within ten million years of the first stellar contractions, the almost-nothingness of interstellar grains had merged into sixty-mile-diameter balls, like the asteroids Ceres and Vesta. Planetary geologists call these country-sized masses planetary embryos, for it was their cataclysmic collisions that formed the rocky planets from Mercury to Mars, as well as, many planetary geologists believe, the cores of the gas-giant planets Jupiter and Saturn and their more distant icy neighbors, Uranus and Neptune. While rocky, metallic objects formed closer to the Sun, it's thought that comets formed out past the Solar System's ice line—the point at which it was cold enough that water in the protoplanetary disk existed as ice, combining with a mix of rocky material to form dirty ice balls.

Astrogeologists estimate that within the first million years of the Sun's ignition, about 4.567 billion years ago, more than a hundred Mars- and Moon-sized planetesimals were jockeying for orbital position within the space that's now the Earth's distance from the Sun. This crowd of early planetary contenders wasn't so much whittled down as it was forced to play together. For between fifty million and one hundred million years, these planetesimals collided, merging through molten collisions into the rocky planets we know today.

What emerged from this great coming together of stardust was a molten mass of a planet. Not a pale-blue dot but a reddish, squirming orb—hot, energized, and angry from the trauma of birth, and yet, as we are testament to today, a planet that somehow cried with the essence of life. But how could a cosmic rock come to life? Rocks from space are once again revealing the secrets of an ancient story.

SAGAN'S DREAM

From a glass-fronted cabinet in one of the NASA-Ames astrochemistry labs, Scott Sandford pulls out an unusual but historic piece of lab glassware: an empty champagne bottle. The bottle's original label is plastered over with a new one, a white piece of paper bearing a scientific graph with a single horizontal line stretching from left to right, interspersed with three sharp, spiking peaks. Those spikes are the reason the bottle is empty. Whatever the champagne's original vintage, on March 14, 2000, Sandford and colleagues emptied it and relabeled it to celebrate a cold cosmic cloud vintage with a wonderful bouquet of amino acids. Like explorers making their mark on newfound land, the scientists tagged the label with their signatures. Among them were the names of Lou Allamandola, the astrochemistry lab's founder and one of Sandford's mentors; Max Bernstein, a former Ames research colleague and now a NASA administrator; and

Sandford himself, his signature directly under the spiking graph. The occasion would have also warranted the handing out of cigars (if smoking were allowed in the lab), for the scientists were celebrating a birth of sorts.

As the current label indicates, the scientists had mixed water, methanol (rubbing alcohol), hydrogen cyanide, and ammonia—a classic foursome of common interstellar molecules—and zapped the mixture with ultraviolet light from a hydrogen lamp in the lab's cryovacuum system. The cryovacuum system's heart is a shiny metallic chamber in which, in a volume smaller than that of a pound-block of butter, the Ames researchers routinely simulated the conditions in cold cosmic clouds—the kind from which solar systems are born. In the lead-up to popping the champagne cork, the scientists had activated the system's vacuum pump to suck out air and drop the internal pressure to less than one-millionth of the Earth's atmospheric pressure at sea level. Meanwhile, the system's cryocooler chilled the system's core to –441°F. When the chamber was completely otherworldly, they'd sprayed in their simple gaseous mix. Before you could say "chilly," the molecules had frozen out on a metallic plate inside the chamber. But the water didn't freeze into the organized crystals associated with ice cubes and snowflakes. Rather, it became haphazard amorphous ice—water molecules so cold they just stopped moving, without time to even get organized. Embedded in that ice were the three other molecules the researchers had injected, and soon the iced-over metallic plate was turned to face the full brunt of an ultraviolet shower from the system's hydrogen lamp, a stand-in for a star. Ultraviolet photons ricocheted into the amorphous ice mixture, shattering molecular bonds and creating molecular fragments and orphaned atoms.

When this cold cosmic chemistry experiment was complete, the cryovacuum system was allowed to warm up to room temperature. As it began to warm, the ice sublimated, going directly from solid to gas; this is the same process that happens to ice cubes in a frost-free freezer that gradually disappear when not used. And there, left as a barely visible residue on the metallic head, were the end products of what astrochemists imagine goes on today in cold molecular clouds—and what went on in the cold molecular cloud from which our Solar System emerged. When Sandford and his colleagues analyzed their astrochemical haul, they detected hundreds of different molecules, most at very low (parts-per-million) concentrations. The new champagne bottle label marks two special molecules they had made: glycine and alanine, two of the most common amino acids, the building blocks of proteins. The NASA-Ames scientists had mixed four simple molecules under conditions that decades ago were unthinkable for chemistry, and they'd produced not just any end products but two molecules that are considered cornerstones of life—and they'd done it by simulating the birthplace of solar systems.

With this experiment and many others, Sandford and his colleagues of the past twenty years have been piecing together one of the greatest current puzzles of the Stardust Revolution: to understand the cosmic origins not of the Solar System's minerals but of its organic molecules. They're creating a detailed molecular family tree that traces our specific carbon ancestry from the molecules of terrestrial life to the stars. The NASA-Ames group is one of the few in the world that's able to tie all the required pieces together: the collection of detailed telescope observations of what's actually out there; laboratory simulations of possible chemistry under diverse cosmic conditions; and hands-on astrogeology, in the form of cometary and meteorite samples, to trace the genealogy of the organic, or carbon-based, molecules that arrive on Earth in carbonaceous chondrite meteorites, comets, and interplanetary dust particles.

In mimicking the coming together of stellar dust grains, water, organic molecules, and starlight, Sandford is realizing the great dream of astrochemist Carl Sagan. Back in the early 1950s, Sagan, then an astronomy undergraduate student at the University of Chicago, sat in on Harold Urey's legendary lectures on planetary formation. At the same time, Urey's graduate student Stanley Miller was conducting the famous primordial-soup experiments that showed that it's possible to make the amino acids central to life from the simplest of molecules, ones thought to have been abundant on the early Earth. For Sagan, Miller's discovery was an inspiration rather than an answer. Smitten by the implications of Miller's result, Sagan decided that his life's mission was to extend Miller's research beyond the Earth, to the cosmos: to understand the cosmic primordial soup. At NASA-Ames, Scott Sandford and colleagues are doing just that. Theirs are the twenty-first-century cosmic versions of Stanley Miller's experiments. Sandford is part of a new generation of astrobiologists who are combining astrochemistry observations, hands-on analysis of extraterrestrial samples, and experimentation to piece together a new story of the origins of life on Earth in a cosmic context. When Scott Sandford first held stardust thirty years ago, he thought it was all about rocks. Today he's a rock hound who's turned prospector for cosmic life.

DNA FROM SPACE

When the Stardust probe returned to Earth from its rendezvous with comet Wild 2, Sandford wasn't waiting to collect bits of rock swept up from the vapor trail of this dirty cosmic ice ball. He was after molecules. What the Stardust researchers confirmed when they parsed Stardust's microscopic payload was that, as suspected from infrared telescopic observations, comets aren't just gigantic, dirty snowballs; they're gigantic, dirty, *organic* snowballs. The molecular evidence swept up by Stardust showed that if you'd been traveling in its wake and had metaphorically lowered your rocket windows, you'd get a very diffuse whiff of something resembling diesel exhaust. The Sun's heat and wind weren't blowing off just bits of cometary dust and water but also the ringed carbon molecules called polycyclic aromatic hydrocarbons—key components of diesel exhaust—as well as a rich stew of other oxygen-, nitrogen-, and carbon-bearing molecules.

What made this discovery all the more interesting was that comets are considered to be the Solar System's ultimate time capsules, agglomerations of the primordial dust, ice, and molecules from which the Solar System formed, preserved since then in a permanent deep freeze out past Neptune, in the Solar System's Kuiper Belt. Thus, like light from distant galaxies that reaches Earth to reveal events from billions of years ago, these cometary samples tell a story not from today but from our planetary beginnings. The Stardust samples also revealed that comets and asteroids are much more similar creatures than previously thought, with overlapping origins. The material collected from Wild 2 revealed that the early Solar System had been a great mixing pot, with material gravitationally churned toward and away from the Sun as if by a mixing spoon. For example, the stardust samples contained a mix of mineral grains, some of which had been melted in the heat from the newborn Sun, and others that had never warmed and thus must have always been in the Solar System's outer reaches. Thus the material from Wild 2 resembled a carbonaceous chondrite, a chip off a dark asteroid, the most primitive and unaltered type of meteorite.

The Stardust samples were the latest in a century and a half of carbon-rich objects from space to get scientists talking about the relationship between carbonaceous chondrites, comets, and life on Earth. In the Stardust Revolution, these streaking cosmic objects have gone from being seen as the age-old harbingers of death and disaster to offering tantalizing clues about the origins of life on Earth. We're at a remarkable stage with each comet sample and carbonaceous chondrite, and now, too, with a bevy of planned sample-return missions to comets and asteroids, adding a new piece to the story of our cosmic molecular origins.

What we've found is that a supermarket-style tour through the organics section of a carbonaceous chondrite is the molecular equivalent of going into outer space and finding carrots, bananas, and broccoli. The soluble organic molecules found in meteorites—those that readily dissolve in water—form life's basic shopping list of ingredients. Carbonaceous chondrites contain not just the simplest organic building blocks, such as methane, ammonia, and formaldehyde, but also the relatively complex suite of molecules that come into play when we make decisions about what we'll have for breakfast. This suite includes proteins and fats, as well as the molecules that make DNA and give us the personality traits that allow us to choose between the sugary donut or the high-fiber bran muffin. According to astrobiologists, carbonaceous chondrites are molecular life kits that arrived filled with the ingredients needed for the emergence of the Earth's first organisms.

And scientists are still learning how to read the cosmic stories held by carbonaceous chondrites. For example, most of what we know about the organic nature of the most famous carbonaceous chondrite, the Murchison meteorite—which crash-landed in Australia on September 28, 1969—has been gleaned only in the past decade, more than thirty years after the meteorite's arrival. Fifty years of technological advances has made a huge difference in the debate over carbonaceous chondrites. Now, rather than arguing about the generic nature of oily carbon materials—as was the case in the 1960s—stardust scientists are able to burrow down to parts-per-billion levels of detail for specific molecules. For the Murchison meteorite's fortieth birthday, a German-led team of researchers gave the meteorite a kind of organics general physical. They didn't search for particular molecules, as previous studies had done, but instead took an overall inventory of any organic molecules they could measure. The result was stunning. All previous research had identified about five hundred different organic molecules in Murchison, but, using a higher-precision chemical inventory method, the research group found more than *fourteen thousand* types of molecules. All this in only several milligrams of material taken from the meteorite's pristine interior. As a result, they estimated that Murchison could contain millions of different organic molecular structures—a cosmic big-box store of organics. Whereas the Earth was previously considered an organics oasis in a barren star field, this study provides a powerful hint that Earth's life-related chemical diversity might in fact be a small subset of an even richer cosmic diversity.

One of the most common soluble organic molecules in carbonaceous chondrites is carboxylic acid. Chemists spot a carboxylic acid when they see a chain of carbon atoms that ends with a familiar carboxyl group: a carbon bonded to an oxygen atom on one side and an oxygen-hydrogen combination on the other. Most of us, though, know carboxylic acid when we taste it. Vinegar, or acetic acid, is the simplest form—just a two-carbon chain, including the carboxyl group, giving vinegar its sour taste. From our first day, we graspingly gulp it down in the form of caprylic acid, an ingredient in breast milk. As kids, many of us survived on it as arachidic acid, a component of peanut butter.

Although the discovery of carboxylic acid doesn't make great news headlines, the discovery of amino acids in carbonaceous chondrites does. These building blocks of proteins are often referred to as shorthand for life. Every muscle in your body—from your heart to the little muscles that move your eyes—owes its structure to long strings of amino acids joined into a peptide chain. These chains provide both protein structure and function. Some proteins are enzymes, the body's chemical matchmakers. Enzymes bring together other molecules and speed up chemical reactions. Without them, the chemical reactions in our bodies wouldn't happen quickly enough to keep us alive, and we'd experience a fast death from chemistry that was too slow. When you see a buff bodybuilder's flexed biceps, you're seeing the cumulative effect of trillions of amino acids. They also bulk up carbonaceous chondrites.

Life on Earth uses twenty different amino acids, each one a variation on a central three-part chemical theme that includes a carbon chain core with a nitrogen-containing component at one end (the amine part) and a carboxylic acid part at the other, hence *amino acid*. But more than eighty amino acids have been isolated from the Murchison meteorite, including eight of the twenty used to build us. Our bodies manufacture some amino acids, but others—"essential" amino acids—we get in our foods. Without them, we're not able to manufacture the full range of proteins that our bodies require. Three of these essential amino acids—leucine, isoleucine, and valine—are all found in the Murchison and other carbonaceous chondrites: you could pop a valine out of Murchison meteorite and exchange it for one in you. Few facts in the Stardust Revolution speak so eloquently to our cosmic ancestry. Our bodies both manufacture and require amino acids identical to those contained in meteorites that, until recently, were at home in the asteroid belt between Mars and Jupiter. Almost half (and probably more, as new carbonaceous chondrite samples appear) of our twenty amino acids are a fingerprint not only of terrestrial life but also of cosmic organic chemistry.

One of the critical achievements of stardust scientists in the past decade has been a conclusive confirmation that these meteoritic amino acids arrived in the meteorites rather than being terrestrial contamination. The evidence for this extraterrestrial origin of amino acids is based on a trio of facts. First, there's a much greater diversity of molecules in carbonaceous chondrites—known as "extraterrestrial analogs"—than on Earth. Many of these molecules are either extremely rare or nonexistent on Earth—as shown by the eighty Murchison amino acids—so they must have come from somewhere else. The second clue is that the meteoritic molecules have distinctly alien isotopic signatures, an indication that they were formed in the ultracold environment of space.

Finally, the most intriguing evidence relates to the fact that some parts of the cosmos are more chemically ambidextrous than here on Earth. Some molecules, including amino acids and many sugars, are chiral—their structure is such that they have naturally forming left-handed and right-handed versions, mirror-image molecules that, just like your hands, can't be superimposed. Even though the atomic contents of the left- and right-handed molecules are exactly the same, the left- or right-handed configuration is pivotal to how, and whether, the amino acid can bond with another molecule. Just as you can't put a right-handed glove on your left hand, you can't mix left- and right-handed amino acids. Amino acids in life on Earth are totally southpaw. We've evolved such that our amino acids are exclusively left-handed, and all our sugars, with which the amino acids bind, are right-handed. However, in carbonaceous chondrites there's a *mix* of both right- and left-handed amino acids, though often with a slight balance in favor of the lefties. This meteoritic evidence for the extraterrestrial formation of amino acids was further supported with the discovery of the amino acid glycine in the material collected by the Stardust mission from comet Wild 2, the first amino acid detected in a comet. Thus, their diversity and isotopic and structural chemistry combine to seal the case that these meteoritic amino acids are indeed otherworldly.

Although the discovery that carbonaceous chondrites are packed with the stuff of proteins has energized the Stardust Revolution, the one-two punch is that they contain not only the building blocks of life but also the blueprint materials. The latest confirmed extraterrestrial organic bounty in carbonaceous chondrites is the presence of nucleobases, the stuff of DNA. Nucleobases are the largest components of nucleotides, the letters of the DNA and RNA alphabets. The four DNA nucleobases, or just bases, in their abbreviated forms are sometimes known as A, G, C, and T, for adenine, guanine, cytosine, and thymine. In RNA, a sister molecule of DNA that acts as an intermediary in the building of proteins, thymine gets bumped in favor of uracil, U.

Nucleobases are the structural and functional foundation of our genetic code, the bits that bind to form the long chains of DNA that are our genetic blueprint. On their own, they carry little information, but put billions of them together, and they become the letters of our genetic alphabet. What meteorites reveal is that this is a cosmic alphabet. Meteorites carry the essence of our genetic code, one that formed abiotically in space before there was life on Earth—before there even was an Earth. In 2008, British astrochemist Zita Martins performed a kind of cosmic genetic forensics. Through stable isotopic fingerprinting—the same kind used to show that meteoritic amino acids have extraterrestrial origins—she showed that the nucleobase uracil that she'd extracted from the Murchison meteorite had formed in space rather than from terrestrial contamination. In 2011, a NASA-led group extended this line of research to show that Murchison and ten other carbonaceous chondrites contain a wide range of extraterrestrial nucleobases, including the core parts of adenine and guanine.

The importance of the discovery of these extraterrestrial nucleobases to our cosmic heritage goes beyond its link to our genetic code. All the nucleobases are formed from either a single or a double ring of carbon atoms, with other atoms branching off these central rings. These cyclic rings of carbon are not only central to our information system; they're also the backbone for energy-transfer molecules and those that perform a suite of other functions. The central molecule in energy transfer in all animals, adenosine triphosphate—more commonly called ATP—is based on a nucleobase-like molecular skeleton. As a final indication of just how important this cosmic carbon-ring framework is to the proper functioning of our daily lives, consider this: the double-carbon-ring-based molecule xanthine was also extracted from the Murchison meteorite. If you chemically accessorize xanthine just a touch with the addition of a couple of simple methyl groups—a carbon and three hydrogen atoms—you get a truly marvelous molecule: caffeine. It's thus quite possible that somewhere out there in the asteroid belt orbiting our Sun is the cosmic equivalent of the jolt that enlivens your morning java.

Most of a carbonaceous chondrite's organic carbon, up to 95 percent, is in the form of a sticky, tarry mass that reminds meteorite researchers of terrestrial kerogen, the decomposed organic matter found in sedimentary rocks. On Earth we search and fight endlessly for the chemical by-products that kerogen produces when it is heated inside the Earth: natural gas and crude oil. Kerogen in meteorites, as on Earth, is made up of an interlocked, complex network of rings of carbon, with a spicing of hydrogen, oxygen, nitrogen, and sulfur. Some researchers think this abundant tarry carbon is the end product of accumulated polycyclic aromatic hydrocarbons, the abundant, large ring-shaped carbon molecules thought to be pumped out by dying stars. But since this glob-like part of carbonaceous chondrites is insoluble, it remains a kind of black hole that holds its individual secrets in an often-impenetrable mass.

In exploring carbonaceous chondrites, stardust scientists have found that they contain all the ingredients of life. Yet "ingredients" doesn't fully capture the scope of molecular diversity found in carbonaceous chondrites and comets. We see an entire molecular system. There are the information molecules in nucleobases; a wide array of structural components, including long chains of carbons and complex carbon rings; there are the nitrogen- and phosphorus-bearing molecules key to protein building and energy transport in cells; and there's a mix of water-soluble and not readily soluble carbon molecules—ones that are ready to react, and others that form a kind of molecular carbon warehouse awaiting the right conditions to release their wealth.

For the scientists at the leading edge of the Stardust Revolution, the bounty of organic molecules in meteorites prompts one key question: What's their direct relationship to life on Earth?

FROM ETERNITY TO HERE

When Apollo mission astronauts stepped, jumped, putted, and drove across the Moon, most television viewers were focused on the astronauts and their experience of being on another world. The first generation of planetary geologists, however, had their eyes on the rocks. The lunar-surface samples returned by the Apollo missions were geologists' first opportunity for lunar "ground truth"—the geologist's term for the clearer picture of reality that comes from actually visiting a locale and collecting samples. One of the key questions geologists asked those Moon rocks was: *How old are you?* The answer, gained through radioactive dating, provided ground truth not just about the Moon but also about a lost period in the Earth's formative history, one that's as much about life as rocks.

The dating of rocks from the various craters that were the Apollo mission landing sites showed that much of the Moon's pockmarked surface is the result of a relatively brief lunar developmental period called the Late Heavy Bombardment. This was a period of several hundred million years following the Earth and Moon systems' initial formation from the protoplanetary disk, when the Solar System was strewn with copious planet-building leftovers: from bits of dust to asteroids to comets and enormous planetesimals hundreds of miles across. It was an era that would bring joy to those who love crash-bang movies. For millions of years, the natal Earth collided with these other bodies, a period of heavy bombardment from Earth's perspective, but if you asked the other drivers, you'd get another answer. On Earth, the evidence of this period has been eroded by tectonic forces, the movement of continental plates. But the crater-pocked surfaces of the Moon, Mars, and Mercury bear testament to these collisions during our Solar System's early, formative years. In retrospect, the Late Heavy Bombardment appears as a logical tailing-off of a bumper-car-like process of collisional accretion that formed the Earth and other rocky planets. Rather than anomalies, cosmic collisions are the nature of Solar System development, and the Late Heavy Bombardment was the tail end—though not the end, since the process continues today—of an essential formative process.

Initially, biologists envisioned this period of intense cosmic collisions as a brake on the emergence of terrestrial life—that just as some primordial reproduction was heating up, the Earth would have been pounded by a planetesimal, asteroid, or comet that vaporized oceans and left the natal Earth cloaked in a cold, choking haze of dust. But Stardust Revolution scientists are rethinking this period, not as one of early life's brutal bombardment but instead as a critical period of seeding the Earth with the cosmic molecules of life.

This new way of seeing was inspired initially by thinking not about life's extraterrestrial start but about its end, by the realization of the role of past asteroid and comet impacts in mass extinctions on Earth. While today the image of a terrified dinosaur looking up at an incoming fireball is the stuff of kindergarten books, our understanding of the role of cosmic impacts on the snuffing out of past life is a remarkably modern one. It wasn't until the early 1960s that geologists generally agreed that many distinctive circular craters found around the world are the remains of ancient cosmic impacts, rather than, for example, ancient volcanic rings. Similarly, the first evidence for an end-of-dinosaur-era impact didn't arrive until 1980, in a paper by Luis Alvarez, who reported the discovery of the K-T boundary—a sixty-five-million-year-old planet-wide geological layer that contains a high level of iridium, a metal rare on Earth but more common in asteroids, and tektites, glassy shards caused by the melting and re-forming of rock during a massive impact. Here was the smoking gun that proved how the end of the dinosaurs came about. The evidence for a cosmic collision's impact on ancient life was sealed in the early 1990s, when the impact was linked to the "crater of doom," the Chicxulub crater, an impact crater more than one hundred and ten miles across, discovered buried beneath the Yucatán Peninsula and dated to the same time as the K-T boundary.

For Carl Sagan, and for Christopher Chyba, his Cornell colleague at the time, this new view of the asteroidal demise of the giant lizards was a clue to an even more ancient story about the role of cosmic collisions in the history of life on Earth. The combination of the awareness of the role of cosmic collisions in the story of life on Earth, the revelation of the Late Heavy Bombardment, and the growing evidence of the carbon-rich content of carbonaceous chondrites all came together to tell another tale. In 1992, four years before his death, Sagan, along with Chyba, proposed a new cosmic-delivery view of the origin of Earth's primordial molecules of life. The Earth didn't form with its thin life-sparking organic layer intact; life's ingredients arrived later via cosmic delivery.

Sagan and Chyba multiplied the higher rates of cosmic collisions during the Late Heavy Bombardment with the organic molecular content of carbonaceous chondrites and cosmic dust particles. They found that, rather than whacking life, the Late Heavy Bombardment delivered as much, if not more, organic material to the young Earth as could have been produced in its atmosphere or oceans through local chemical reactions. Sagan and Chyba had extended the famous Miller-Urey experiment into a cosmic context. They showed that the chemistry of life wasn't dependent on starting on Earth, but rather that the early Earth was literally seeded with life's core ingredients.

They concluded that most of this carbon arrived not with a meteoritic bang but with a steady organic dusting, molecules drifting down as part of cosmic dust over millions of years, transforming the early oceans into a rich organic primordial soup, a broth of life.

But could a gradual, "green" rain over several hundred million years really deliver enough organic material to make all the fish, birds, trees, fungi, microbes, and, of course, people that make up terrestrial life? Subsequent studies have shown that the amount of organic material that splashed down on Earth during the Late Heavy Bombardment was actually many times more than currently on the Earth's surface; that our planet has in fact lost carbon to space in the form of volatile carbon molecules such as carbon monoxide or carbon dioxide, or that this organic material has been buried deep below the Earth's surface by geological processes.

The cosmic, postnatal delivery of Earth's organic layer is also supported by the fact that many astrobiologists think the evidence shows that a significant portion of the water that covers three-quarters of the Earth's surface similarly arrived via cosmic delivery in the form of water-rich asteroids and comets. The exact origin, timing, and delivery mechanism for Earth's water is still a hotly debated topic, with some researchers pointing to the possibility of off-gassing from water-rich molten rock, and others pointing to new evidence that some asteroids are much wetter than previously thought and that some comets bear the same isotopic signature as the water in Earth's oceans.

"Everyone used to talk about the origin of life as if it all happened on Earth," says astrobiologist Scott Sandford. "A lot more stuff falls out of the sky. That could have played a big role. There's been a tendency for people to divide themselves into camps" on the origin of the Earth's carbon molecules, "but in reality that's probably crazy. It may well be that the net story for Earth could have been a very complicated and confused one. It may have been that when early life needed amino acids, it found them in reservoirs that contained both an extraterrestrial and terrestrial component. They were being made in both. Probably, early life needed some materials that by and large were heavily dominated by what was falling out of the sky, and other components that were largely being made terrestrially. In the end, life probably took whatever it needed and whatever it could find."

The first tentative fossil and biochemical evidence of bacterial life on our planet appears in ancient ocean sediments in Greenland formed during the latter part of the Late Heavy Bombardment more than 3,850 million years ago. Those first marine cells contained a remarkably recent cosmic heritage. Their bacterial bodies contained molecules that only hundreds of millions of years before, or less, had been the stuff of a star-birthing cosmic cloud. The carbon had spun around its nascent star, clumped into dust, maybe into an asteroid or comet, and eventually arrived on the surface of a young world. Now as carbon-based molecules, they were reproducing, adding their own story to that of this cosmic carbon, shaping it into new proteins, sugars, and eventually long threads of the genetic molecules that would carry this cosmic molecular heritage into the creation of a new world.

TRACING OUR COSMIC CARBON ANCESTRY

With the knowledge of organic molecules in carbonaceous chondrites and comets and their cosmic delivery to Earth, one of the most intriguing, ongoing mysteries of the Stardust Revolution is the question of the cosmic origin of these carbon-based molecules. While fifty years ago astronomers didn't even think there was cosmic chemistry, today, in the Stardust Revolution, there are so many potential locales and chemical pathways for cosmic organic chemistry that the heated discussions are about which one, or ones, leads to us. In some situations, it's a case of the blind astrochemists and the cosmic molecular elephant. Different researchers literally see different evidence: radio astronomers see the spectroscopic signatures of molecules as gases around old stars; infrared astronomers try to deduce the nature of molecules frozen in ices and large clumps of carbon dust in interstellar space; and astrogeologists try to trace our carbon ancestry backward by analyzing the nature of organic molecules in meteorites. Is the molecule whose spectroscopic fingerprint is seen by a radio astronomer streaming away from a billowing red giant star the same molecule that dissolves out of a carbonaceous chondrite on Earth? Similarly, there's such a variety of forms of cosmic carbon-based molecules—from the biggest multi-ring aromatic molecules to the enormously abundant but simple carbon monoxide—that trying to trace our cosmic carbon ancestry is a case of looking at a vast genealogical database and having to work, and argue, just to know where to start looking for any particular genealogical lineage.

We're at the start of tracing our detailed cosmic carbon ancestry, and with each new observation, experiment, or meteorite sample, we are adding a little more color to the story—a process akin to knowing where a great-great-grandmother was born or that an even more distant relation was the one who'd made the great immigrant's journey from a distant land across the sea. Tracing our carbon ancestry has far-reaching implications for both understanding our own cosmic origins and the likelihood of life elsewhere. For example, if most key organic chemistry occurs around newborn stars, rather than all newborn solar systems being stocked with complex organic molecules from earlier cosmic processes, this fact would set boundaries on the likelihood of other organic-rich planets.

The weight of evidence that's emerging in the polyvalent effort to trace our cosmic carbon heritage is that the physical origin of the molecules that rain down on Earth largely predates the formation of the Earth. When we trace the genealogical heritage of the building blocks of life found in carbonaceous chondrites and meteorites, we find ourselves led back not to some little organically rich pond on the primordial Earth but to the cold molecular cloud from which our Solar System, and hundreds of others, emerged. What's clear is that the organic molecules that arrive on Earth were formed in a variety of astrophysical environments, from around dying stars, to within cold molecular clouds, and to within asteroids and comets.

Our carbon story begins with the carbon atoms that are born in the hearts of dying red giant stars from the fusion of three atoms of helium. One by one, the carbon atoms are dredged up in great convection currents to the star's outer atmosphere and blown outward on stellar winds to begin a journey that will involve a complex dance of molecular breaking, recycling, marriage, and familial bonding. Some astrochemists think that the vast bulk of the first generation of carbon-based molecules formed around these dying red giant stars are ripped apart in less than a century, the equivalent of a cosmic blink, by the intense ultraviolet radiation from the white dwarf star, the red giant's stellar remnant. However, University of Arizona astrochemist Lucy Ziurys sees evidence of abundant molecular survival around old stars, with molecules shielded by dust, water vapor, and other molecules. "It appears that molecules are surviving everywhere in old planetary nebulae," the molecular remnants of dissipated red giant stars, Ziurys tells me. "We don't wipe the chemical slate clean with each stage. The origin of organic material on Earth—the chemical components that make up you and me—probably came from interstellar space. So one can say that life's origins really begin in the chemistry around [these old stars]."

One way or the other, big or small, after millions of years in interstellar space, carbon-based molecules eventually find themselves swept up into the cold, dense molecular clouds from which solar systems emerge. Through laboratory simulations, telescopic observations, and the analysis of organic materials in meteorites, what's become clear in the past decade is that these molecular clouds, and particularly the densest clumps from which stars emerge, are also the birthplace of organic molecules. Stars and complex carbon molecules fuel one another's creation.

For example, astrochemists have described what could be called cold cosmic flagpole chemistry in these dense molecular clouds. In this process, molecules form as a dance between ice-covered micrometer-sized dust and the small molecules that stick to these grains, like a child's tongue to a frozen flagpole. It's thought that methanol—the simplest alcohol, also called wood alcohol, the fuel used in drag-racing cars—is produced when carbon dioxide gas freezes onto a cold dust grain and chemically reacts with hydrogen molecules there. In a similar process, it's thought that the methanol later reacts to form formaldehyde, a molecule that readily joins in chains to form many larger, complex organic molecules. Thus the molecule that on Earth we associate with preserving corpses may, on a cosmic scale, be a key link in the chain from stars to life.

The analysis of the organic materials found in meteorites confirms that some of this material predates the Sun. When astrochemists took the isotopic fingerprints of distinctive, globule-shaped clumps of organic molecules in the Tagish Lake carbonaceous chondrite—whose charcoal-like chunks crash-landed in the Yukon, Canada, in January 2000—they found they all bore the same label: made in a cold cosmic cloud. The meteoritic molecules' isotopic signatures revealed that they were formed at cosmic deep-freeze temperatures, either in the cold molecular cloud from which the Sun formed or at the frigid outskirts of the solar nebula, after which they were incorporated into the asteroid that spawned the Tagish Lake meteorite. Similarly, the amino acids in the Murchison meteorite also have isotopic signatures that peg them as having formed in a cold molecular cloud.

Indeed, at NASA-Ames, Scott Sandford and his colleagues have verified that supercold is no barrier to abundant organic chemical synthesis. In their cryovacuum system, they've cooked up thousands of organic molecules, only a small fraction of which have been identified. Those positively fingerprinted include the amino acids glycine and alanine; the nucleobases uracil and cytosine; quinones, an important class of biologically active molecules, including 1,4-naphthoquinone, which is the core of vitamin K; and a flood of simple organic molecules such as formaldehyde, methanol, and methane.

"It turns out that a lot of our chemistry happens during the warm-up," Sandford explains.

> So probably what happens is you shoot the photons in, they hit all these molecules, and they break a lot of bonds. So you break the

> water into OH and H, and you break the methanol into CH_3 and OH and so on. These are all highly reactive molecular species. They would love to do chemistry. But they can't because they are frozen in the ice. So they just sit there with all this chemical potential. But the minute you start to warm up and the ice can rearrange, then all these things find each other and they start reacting like crazy. It's a very bizarre chemistry, not the kind you do in freshman chem lab.

Yet it could well be the chemistry that occurs around newborn stars. The heat from a gestational star's friction, and then initial nuclear reactions, warms up and softens icy grains in which previously formed molecules gain the energy to move and interact, forming more complex, larger molecules. This warming, melting, and molecular moving and mixing also occurs in asteroids and comets, turning them into orbiting chemistry labs, each one on its own chemical journey. There's evidence from the Tagish Lake meteorite that, while some of its organic bulk was formed in cold molecular clouds, other molecules, including some amino acids, were formed in the asteroid itself from common precursor molecules during periods in which the asteroid contained liquid water.

We know that pre-solar grains of stardust such as nano-diamonds and silicon carbide survive the journey from their birth in a long-ago star's atmosphere to modern-day Earth. Is it possible that Earth is also pelted with *organic stardust*, pristine molecules that formed around dying stars and that have remained intact across time and space until they crash-land on our pale-blue dot? Some stardust researchers believe that at least part of this complex insoluble organic material in carbonaceous chondrites comes directly from stars. The evidence, according to the foremost proponent of this view, Hong Kong–based stardust scientist Sun Kwok, is in the enigmatic spectroscopic infrared fingerprints of molecules around dying stars. The spectroscopic fingerprints of dying red giant stars contain a large swath of unidentified infrared emission bands. There's something there; the question is: What? Kwok believes that the emissions are from nanoparticle-sized particles of complex mixes of rings and chains of large carbon-based molecules. It's a structure very similar to that identified in the interlocking weave of branched, ringed carbon-based molecules extracted from the Murchison meteorite. If this view is correct, it means that organic stardust older than the Earth, and largely unchanged over billions of years, rains down on our planet. Thus, in some cases, our carbon ancestry is perhaps but one degree of separation, a single molecular generation, away from its birth star. In this case, when we talk of being stardust, the lineage is stunningly direct.

For astrobiologists like Scott Sandford, the dream today isn't to understand the cosmic context of just life on Earth but of life on planets in general. "The universe seems to be hardwired to generate real molecular diversity, including some of these astrobiologically relevant molecules," he says. "So, in a sense, the universe is a big organic chemist. Not a very efficient one, but the beakers are so big, and the time scales are so long, that in the end you can produce an awful lot of material . . . I think you can pretty much count on the likelihood that when a new star forms, if it's got planets around it, the surfaces of those planets will in fact have a complex organic mix dumped on them."

Whether or not we know if this is what happened on Earth will largely depend on finding another living planet, a point of cosmic comparison, the great dream of the Stardust Revolution: to find an alien Earth.

CHAPTER 8

OTHER WORLDS

The province of the student of astrophysics may be said to end with an understanding of the production of a planet like the Earth.

—George Ellery Hale, *The Study of Stellar Evolution*, 1908

NEW FRONTIERS

At first glance, Michel Mayor could easily be one of the late-summer tourists stopping at Wyoming's Jackson Lake Lodge to get a bite and enjoy the lodge's unparalleled view of the gray, angular Teton Mountain Range rising precipitously from the flat plain ten miles to the west. The sixty-nine-year-old Mayor's relaxed casualness—red sweater, beige slacks, Velcro®-buckled sandals over socks—isn't that of a tourist but of the emeritus astronomer. For Mayor, the astronomy conference he's here to attend is a gathering of the clan, a homecoming of sorts for astronomers from around the world—with all its rivalries, friendships, and black sheep. Mayor is the clan's great elder, the one who finally broached a seemingly impassable frontier and opened the way to new, unimagined worlds. While tourists snap pictures of the Tetons, there—sitting in an upholstered, wingbacked chair, head down while working on his laptop in the Jackson Lake Lodge atrium—is a Galileo of our times.

The setting is a gustatory purgatory for the Geneva-based Swiss astronomer. The lodge's diner, with its classic linoleum-topped counter, is a testament to Americans' penchant for speed over aesthetics. But the organizers of the Extreme Solar Systems II conference chose the locale for its symbolism, not its food. Wyoming is synonymous with the US frontier of western settlement, a place where many people still wear cowboy boots and Stetsons®, and there's even an outside chance they need both for working on the range. For the twenty-first-century cosmic explorers meeting here for the conference, few places evoke such a strong sense of both exploration and the Earth as a dynamic planet. Jackson Hole gets its name from the fact that it's sinking. The Teton Mountains are North America's youngest, most active mountain range. With each jarring earthquake, they strain skyward on one side of a geological fault, while on the other side, gravity pulls down the slab of the Earth's crust that is Jackson Hole. This geological activity is even more pronounced about seventy miles north, up the road in Yellowstone National Park, where each year, millions of visitors arrive in geological pilgrimage to see Old Faithful, a geyser whose hot, sulfurous eruptions give vent to the unseen tumult of forces below the Earth's surface.

This is a land of emergent possibility that speaks to Mayor's career. The astronomer has spent a lifetime painstakingly exploring and mapping the hidden crevices of the cosmos, probing deeper and deeper into the unknown, revealing a universe many thought unattainable. He hadn't planned it this way; as a master's student in Geneva in the late 1960s, he studied theoretical particle physics, a field in which a single atom holds the equivalent of a cosmos of activity. One day, he and a friend stopped in front of a bulletin board that announced two PhD research opportunities, one in statistical mechanics—a branch of quantum mechanics—and the other in astrophysics at the Geneva Observatory. His friend expressed interest in the first; Mayor settled for the second.

"I had no special interest in astronomy," Mayor recalls. "I have exactly the same interest for a lot of domains of science." Add to this deep, enduring curiosity the qualities of patience and persistence. With these attributes, Mayor began what's now a five-decade career watching and measuring how stars and planets gravitationally tug on each other. When he began in the early 1970s, nine planets were known in the entire universe: those of our Solar System, from scorched, rocky Mercury out to dim, icy Pluto. But in 1995, from an observatory in Southern France, Mayor and his graduate student Didier Queloz ended millennia of speculation and philosophizing: they discovered the first planet revolving around a star other than our Sun—an exoplanet. It was only one more planet in the cosmos—about half Jupiter's bulk, orbiting its star in just over four days—but the exoplanet 51 Peg b represented a whole new universe.

In fulfilling an ancient dream, Mayor and Queloz opened the door to a modern one. The search for other solar systems and ultimately another living planet is the beating heart of the Stardust Revolution. This quest is about much more than finding extraterrestrial cosmic cousins with whom to talk or visit, or the joy of turning science-fiction visions of alien worlds into reality. Without planets revolving around other stars, the Copernican Revolution was a philosophical dead end. There might be billions of galaxies and trillions of stars, but as far as concrete evidence went, the only known planetary system was ours. Earth wasn't the geographic center of the cosmos, but science fiction aside, it appeared to be the planetary and biological center, and a unique one at that.

For the scientists of the Stardust Revolution, the absence of other planetary systems presented the central conundrum in developing a natural history of cosmic evolution. Astronomers could study the natural history of everything from galaxies on down to stars, molecules, and even atoms. Yet when it came to planets, they hit a wall or, rather, a gap. Without other planets, the cosmic story—*our* cosmic story—didn't really make sense. Here was a universe in which common forces and processes—from gravity to the nuclear processes in stars to astrochemistry—were clearly visible and responsible for the cosmos that astronomers observed, but this physical continuity fell into a black hole when it came to other solar systems. Did the cosmos really behave one way around our Sun and differently everywhere else? Were we, as some leading planetary geologists posited in

2000, a rare Earth, a planet with diverse life that might be "utterly unique . . . in the visible universe"? There was an intellectual chasm at the core of the Stardust Revolution.

It was a gap that was as much about biology as about astrophysics. Without other planets, there was no way of truly understanding our cosmic origins. Evolutionary theory is grounded in comparative biology. Darwin's evolutionary insights came from famously comparing the beaks of different finches, comparing animals on one continent with those on another, and comparing the fossil skeletons of ancient animals with the skeletons of animals walking and swimming during his day. Stardust scientists could compare interstellar molecules and stellar energy output across the Milky Way, but astronomy and evolution could never truly meet without the existence of other worlds with which they could compare biological notes. Astrobiologists needed comparative exoplanetology if they were ever going to deepen their knowledge of our astrophysical heritage.

Mayor and Queloz's historic exoplanet discovery now seems like it took place long ago, like the sepia-toned images of the old West that adorn the Jackson Lake Lodge's walls. In less than two decades from the discovery of the first exoplanet, a new cosmos has emerged. We're no longer alone—at least not when it comes to solar systems; and astronomers already know of hundreds of others. On the conference's first day, Mayor announced his research group's discovery of more than fifty new exoplanets.

But for all the palpable excitement and intense buzz at the Extreme Solar Systems II conference about the number and nature of alien worlds, it was the absence of two pioneering exoplanet hunters that bore testament to just how desperately, dirt-scrabble-hard it had been to cross into this cosmic frontier of alien worlds—and what might lie ahead.

A NEW VISION

Bruce Campbell and Gordon Walker met for the first time in September 1969 at the University of British Columbia in Vancouver, Canada, as classes resumed for the fall session. As Campbell, a twenty-one-year-old, third-year engineering student, walked into his first astronomy class, at the front of the UBC lecture hall was Walker, a Cambridge-trained thirty-three-year-old recent immigrant from Scotland, in his first year as a professor. The two men quickly found they shared a common passion: they were astronomy tech geeks, drawn together by a love of and fascination for building better and faster telescopic equipment. Campbell scored the top mark in Walker's class, Astronomical Measurements, graduated from UBC in 1971, and returned to work with Walker as a postdoctoral researcher in 1976, after earning his PhD at the University of Toronto.

Campbell returned because Walker was, in the *Star Wars* movie jargon of the day, a Jedi knight of stellar spectroscopy in his ability to sift starlight for its secrets. Campbell was Walker's Luke Skywalker—brilliant, talented, and ambitious. Their technical curiosity and prowess led them to realize they could create a radically better way to study starlight, one so accurate they'd be able to see something no one else ever had: the slight gravitational tug of a distant planet on its star.

Although we're used to thinking of stars as fixed points of light in the night sky, stars move. A lot. They expand and contract due to their thermonuclear nature, boiling up in some spots and shooting out arcing fountains of plasma. Stars, like planets, rotate on their axis, a rotation that on Earth creates day and night. And, just like planets, stars have an orbit. Although we think of planets orbiting a stationary star, the star and its surrounding planets actually orbit a common center of mass. Much like when a large adult and a small child balance on a teeter-totter—to make it work, the adult sits very close to the center of mass, and the child sits farther away—the "parent" star is very close to the center of mass, giving it an orbit not much larger than its own diameter. It's this to-and-fro movement that astronomers measure with astrometry—measuring a star's back-and-forth movement across the sky over the course of many nights.

However, there's another less obvious way of seeing a star's movement. Rather than trying to measure a star's side-to-side movements, it's now easier to measure its movements toward and away from the Earth. The secret is in its light. Spectroscopy can be used to see not just what stars are made of but also how they move. Measuring a star's speed around its orbit depends on Doppler spectroscopy, a 150-year-old standard of astronomy. Doppler spectroscopy had already made its claim to cosmological fame in the red-shifted galaxies that Edwin Hubble observed, the red shift indicating that these spirals of stars were speeding away from the Earth and from one another—that the cosmos is expanding.

The Doppler method is based on the same wave behavior that occurs when the pitch of a siren changes as an ambulance speeds toward us, passes us, and then goes away. In the case of stars, astronomers look for changes in the pitch, or frequency, of starlight to measure a star's speed toward or away from us, using a technique called radial velocity. It's colloquially known as "the wobble method" because, plotted on the axis of a graph, the star's speed varies, creating a wave, or wobble, in the curve as the star appears to move slightly faster when approaching the observer and slower when moving away. The bigger the wobble, the more massive or closer the exoplanet is to its star.

In 1952, Russian American astronomer Otto Struve—whose great-grandfather Friedrich had first calculated the interstellar extinction of starlight a century earlier—predicted that the radial-velocity method could be used to detect exoplanets. In the 1970s, the technique was being widely used to detect stellar binaries—pairs of stars in orbit around one another, each gravitationally altering the other's speed, though one star might be too faint for anyone to observe its light. But these spectroscopists were working with errors of four hundred meters per second, margins far too large to detect the slight stellar speed changes induced by a planet.

Campbell and Walker realized they could improve the radial

velocity method by creating a better light ruler. The key to measuring the red shift is to have a reference against which to measure changes in a star's light fingerprint. In the early 1970s, Walker's research group developed a technique to use the light fingerprint of water in the Earth's atmosphere as the ruler, significantly boosting the accuracy of their measurements. But it turned out that this technique was like using a rubber ruler, because atmospheric changes smudged water's spectral fingerprint. Campbell's stroke of genius was to turn back to the technique that Gustav Kirchhoff had used to determine the nature of the Fraunhofer lines. Campbell envisioned that by passing starlight through a known gas placed in front of the spectrometer, the gas's distinctive spectral fingerprint would act as a rigid, finely grained ruler against which to measure changes in starlight.

The key was to choose a gas with regularly spaced spectral lines. Molecular spectroscopy expert Gerhard Herzberg—who'd arrived in Canada as a refugee from the Nazis with five dollars and his beloved spectroscopy equipment—said he knew just the right gas for the job: hydrogen fluoride. Campbell and Walker agreed. Hydrogen fluoride produces a spectrum of evenly spaced absorption valleys, like the tines on a comb across the spectrum. The gas's only drawback is that it's highly toxic and corrosive; mixed with the slightest amount of moisture, such as is found in the lining of the respiratory tract, it becomes a powerful, deadly acid. Campbell and Walker developed a clear, sealed, inert metallic cell with sapphire windows (hydrogen fluoride dissolves glass) to hold the gas that could then be placed in front of the spectrometer. They were concerned that they might not survive to tell the tale of their first test of the system, but the hydrogen fluoride cell worked perfectly, providing an unprecedented level of accuracy. For the first time in history, the velocities of stars could be measured to plus or minus ten meters (approximately thirty-three feet) a second, about the speed of an Olympic sprinter. Thus a telescope and spectrometer could be used (like a police speed-radar gun) to clock a distant star's velocity.

A radial-velocity accuracy of ten meters a second was the magic number for exoplanet hunting. Astronomers knew that Jupiter's massive gravitational tug causes our Sun to speed up and slow down by about twelve meters per second. Thus, Campbell and Walker's technique would spot the wobbles induced by Jupiter-sized exoplanets on stars elsewhere in the Milky Way—if they were out there. "I heard myself say to Bruce, 'We could start looking for planets,'" Walker recalls. "I don't know where the idea came from. That's how these things happen in science—suddenly a light turns on. It's the art of the possible."

Walker and Campbell were moving into an astronomical minefield. If other worlds were out there—which many astronomers believed was the case—they were devilishly hard, perhaps impossible, to detect, judging from a half century of experience. During the five decades from World War II to 1995, numerous exoplanets were thought to have been discovered. In the equivalent of a Harvard Business School case study of the vagaries of scientific discovery—of turning years of research into a hard-won accepted truth—this half century was marked by flattened hopes, faulty technologies and calculations, and off-base news reporting. Headlines with messages like "First Planet Found outside Our Solar System!" appeared in newspapers dozens of times over this period, at least twice in the *New York Times* and once on its front page. It wasn't the reporters who were getting it wrong; it was the astronomers.

The most dramatic and high-profile exoplanet discovery claim—the great cautionary tale among exoplanet hunters—is that of a putative Jupiter-like planet around Barnard's Star, which was heralded by American astronomer Peter van de Kamp in April 1963. He'd hardly rushed to the conclusion. His assertion was based on a painstaking quarter century of careful observations, after which van de Kamp's planet made it into some astronomy textbooks as the first exoplanet. A decade later, van de Kamp's planet disappeared. A fellow astronomer noted that the measurements recording the movement of Barnard's Star on the sky—the basis for concluding there was an unseen planet tugging at its star—were almost identical to the movements of another star. While it was improbable that both stars had absolutely identical movements, the astronomer noted that the apparent shifts in position coincided with equipment upgrades in 1949 and 1957 to the telescope van de Kamp had used. The shocking conclusion was that van de Kamp had recorded the movement not of the star but that of the telescope itself.

In another case of mistaken exoplanet identity, a venerable American astronomer arrived at what was planned to be the announcement of an alien world he'd found; unfortunately, he was forced to announce that what he thought was a planet was in truth a mathematical error. He was nonetheless hailed by his colleagues for having been brave enough to break the bad news himself. And it wasn't only human error that produced supposed exoplanets. The most remarkable coincidental erroneous exoplanet spotting occurred on December 11, 1984, when two American research groups simultaneously raced to announce the detection of an object orbiting a distant star. Ostensibly, the detection of something by two competing groups would encourage any poker player to bet in favor of the object's existence. But on closer inspection, there wasn't anything gravitationally tugging at the star. The postmortem analysis concluded that, somehow, subtle systematic effects of the Earth's atmosphere had similarly skewed both observations. When van de Kamp published his last book, *Dark Companions of Stars*, in 1986, the frontispiece included the biblical inscription "Blessed are they that have not seen, and yet have believed." It was a fitting testament to the deep challenge faced by those who hoped to fulfill the ancient quest of beholding alien worlds.

Walker and Campbell knew they faced a formidable challenge, but they had a plan. After a trial run with the Doppler spectroscopy system on the telescope at the Dominion Astrophysical Observatory, north of Victoria, British Columbia, in 1980, Campbell installed their star speed system on the new, much larger Canada-France-Hawaii Telescope atop the dormant volcano Mauna Kea in Hawaii. They were fortunate to get telescope time, leveraged as it was by Walker's stature in the astronomy community rather than by great enthusiasm among the community for their program. "It is quite hard nowadays to realize the atmosphere of skepticism and indifference in the 1980s to proposed searches for exoplanets," Walker wrote in chronicling his exoplanet search.

Figure 8.1. Pioneering exoplanet astronomer Bruce Campbell inspects spectroscopic equipment at the Canada-France-Hawaii Telescope in 1980. *Image courtesy of Canada-France-Hawaii Telescope.*

Walker and Campbell's exoplanet hunt plan was straightforward: they assumed that other solar systems existed and were like ours; Jupiter-like exoplanets would take about twelve years to orbit their star, just as Jupiter does. Walker, Campbell, and University of Victoria astronomer Stephenson Yang began a decade-long search of twenty-six stars, looking for Jupiter-sized exoplanets—ones large enough, they reasoned, to tug at their stars to a degree visible with the scientists' telescope. Looking patiently at more than two dozen stars and using the best spectroscopic tools in the world, they thought that if there were Jupiter-sized planets around other stars, they'd find at least one. Three or four times a year, one or all members of the team spent several nights atop Mauna Kea, fourteen thousand feet above the Pacific Ocean, searching for other worlds. They'd start just after dusk and work through the night, wearing winter parkas and enduring altitude-induced headaches and dehydration until dawn's light drowned out the stars.

"It was deadly," Walker says of the psychological challenge of searching for exoplanets no one knew existed or that could even be found. "We never knew from one night to the next if we were observing anything. We could only tell when the data were reduced, about once a year, whether we were seeing trends." The work truly was potentially deadly. Working with altitude-induced clouded thinking, the astronomers had to heat the hydrogen fluoride gas to just over 200°F and repeatedly mount and dismount the container. In case it ruptured, they carried gas masks and hydrogen fluoride antidote gel for treating external burns.

By 1987, the astronomers believed they were seeing something no Earthling had ever seen before—stellar wobbles caused by orbiting planets. That summer, at a press conference at the annual meeting of the American Astronomical Society in Vancouver, Campbell announced their preliminary results: a half-dozen stellar wobbles indicative of possible exoplanets. One star's motion was particularly intriguing: that of Gamma Cephei.

It was a fitting star for the Canadian planet hunters. A bright star forty-five light-years away, Gamma Cephei is always visible from Canada in the night sky, shining near Polaris, the pole star. Speaking in a charged, paparazzi-esque atmosphere to a group including the astronomers who packed the press conference room, Campbell described how, from their detailed measurements, he and his associates had seen that Gamma Cephei had a periodic 2.5-year-long wobble. As viewed from Earth, the star moves toward us and then away over a 2.5-year cycle—evidence, they thought, of an exoplanet gravitationally tugging at the star.

The *New York Times* headline read "Planets outside Solar System Hinted." Campbell recalls, "They were calling us planet hunters. We were on the track." His professional colleagues weren't as impressed. In an arena of supercharged skepticism, Campbell's tentative announcement generated more doubt than accolades. An astronomer quoted in the *New York Times* article said, "I probably won't call it a planet until I can get out and walk on the surface of it." No researchers attempted to confirm the results; at the time, none could. Walker says the press conference, ironically, was more like the beginning of the end rather than a high point, increasing the pressure to deliver results from a project that was, by its nature, deeply long-term. Two years later, Campbell, Walker, and Yang still had nothing conclusive, and, worse, Campbell didn't have a secure job. A Vancouver native, he was determined to stay in the area, but a decade of effort hadn't earned Campbell a permanent position at UBC, the University of Victoria, or the National Research Council's Victoria-based Herzberg Institute of Astrophysics. In Walker's view, Campbell's not getting a job "had a lot to do with the fact that planet searching was still not above the radar. . . . It all started to unravel."

Campbell alienated those who controlled Canada's professional astronomy appointments by publicly bemoaning the state of astronomy funding in Canada. Increasingly frustrated, disheartened, and distanced from the academic community, Campbell took a final, renegade approach with the help of an influential supporter—the guru of astronomy popularizing, Carl Sagan. "I remember telling Carl Sagan, we think we're going to be able to do this," recounts Campbell, who met Sagan in the late 1980s while the two were on a University of Toronto panel on the search for extraterrestrial life. "He was somewhat incredulous at first, but then I convinced him, and he became a great friend and supporter."

Buoyed by Sagan's endorsement, Campbell raised $125,000 in private funding to support an endowed planet-hunting chair at UVic. However, the Canadian government's university research-funding guidelines at the time specified that only tenure-track professors could receive funding. This disqualified Campbell, an adjunct professor, from receiving the necessary matching federal funding for the position. By 1991, at forty-two years of age, a salary and family stability were more important to Campbell than unlocking the secrets of the stars. "It had been so frustrating to try to secure some sort of position, and then to try to set up an endowed chair, that when it all came to naught, I decided to walk away," says Campbell. When he did, he went supernova. In a final burst of anger, and in a major breach of research etiquette, Campbell erased his computer hard drives, deleting a decade's worth of compiled and analyzed data. After the deep uncertainty of looking for other worlds, he turned his back on the stars for good and took up a career dealing with one of life's great certainties: taxes. "It was the advent of electronic filing of tax returns in Canada that got me involved," says Campbell, who started work as a personal tax consultant in 1993. "Many accountants were hesitant to get into computers, and I pretty much knew how computers worked, so I had the edge there."

Stunned and bruised by the loss of his younger colleague, Walker was left to lead the conclusion of the exoplanet project. It took almost a year of painstaking work for Stephenson Yang—fortunately by then holding the additional position of UVic's computer systems manager—and UVic colleague Alan Irwin to recover the original data and reanalyze it. In 1992, having reassessed the data, Walker came to a conclusion about the most promising of the possible exoplanets, the one around Gamma Cephei. In a scientific paper, he and several coauthors concluded that the star's 2.5-year wobble was probably due to its own cyclical expansion and contraction, giving the appearance that the star was in fact moving toward and away from us. As the star expanded, it would appear to move toward the Earth, and as it contracted, it appeared to move away. There was no invisible exoplanet, just a turbulent star. Nobody disputed the conclusion, except Walker himself. In fact, he'd quietly agonized over whether the wavy line in the data was in fact a star revealing the presence of another mysterious world.

"I had written the paper as it being a planet," he says. He was sitting in his office when a recently arrived postdoctoral student, Jaymie Matthews (now a UBC professor), entered and looked at the data. Matthews pointed out that the supposed planet's two-and-a-half-year orbital period coincided with what appeared to be periods of the star's heightened surface activity. "I think Jaymie had a very valid point," says Walker, who deferred to the dominant view and rewrote his paper, adding a question mark to the title and saying that what might be an exoplanet might also be stellar rumblings. He'd been right the first time. Canada's planet hunters had found their prize. Yet it wasn't until 2003, after assessing almost twenty years' worth of data, that exoplanet hunters finally concluded definitively that every 906 days—about two and half years, as Walker had calculated—a Jupiter-sized planet completes its orbit around Gamma Cephei. The final vindication of the detection of the planet Gamma Cephei b for Walker "was like a Eureka moment in tortuously slow motion."

David Charbonneau, a leading exoplanet hunter at the Harvard-Smithsonian Center for Astrophysics, says that Walker's experience is a testament to the excruciating nature of the search to discover new worlds. "We like to imagine that the scientists looking through the microscope or telescope see something and then they know that this is the thing they've been looking for and it's just a matter of getting the news out," says Charbonneau. "But the point is that when you're actually involved in a true discovery, it's a very uncomfortable process, because you really don't believe that this is the thing you saw, and you want to make absolutely sure you're not being confused by some spurious signal and you start to question the data yourself."

Within months of Walker's about-face, Polish American astronomer Alexander Wolszczan announced the discovery of two Earth-sized objects around a pulsar, the radio wave emitting the dense remnant of a supernova. This was completely unexpected—astronomers are still uncertain as to how planets survive or result from a star's detonation—and these objects were the first planet-like objects found outside our Solar System. However, in the scheme of understanding planet formation around stars and the search for possible Earth-like exoplanets, these heavily radiated cinders were space oddities rather than guiding lights.

In June 1995, at an astronomy conference in Calgary, Alberta, Gordon Walker told the assembled star watchers that if other Jupiter-sized planets orbited distant stars, they were far less abundant—and much harder to find—than had been assumed. In the audience sat a man who would have loved to speak his piece, but the timing wasn't right. A year earlier, in April 1994, Swiss astronomer Michel Mayor had started his own exoplanet hunt. For Mayor, it was a natural progression. With his graduate student Antoine Duquennoy, he'd recently completed a landmark census of relatively nearby Sun-like stars and had discovered, using radial velocity measurements, that our naked eye deceives us when we look into the night sky. About 60 percent of all Sun-like stars are binaries—two stars in an orbital dance. Mayor was aware of Campbell and Walker's spectroscopic work—he'd met Campbell and talked spectroscopic shop at the first bioastronomy conference in France in 1990. Having conducted the definitive survey of stars orbiting one another, Mayor set his sights on using radial velocity to search for smaller orbiting objects, such as the hypothesized yet elusive brown dwarf—an object a smidgeon too small to ignite as a star—or exoplanets. For the search, he helped build a new spectrometer at France's Haute-Provence Observatory, which used new state-of-the-art digital technologies to achieve a stellar-wobble precision of just fifteen meters (about forty-nine feet) per second. Mayor was in the exoplanet game.

It didn't take long. Relying on his stellar binary work, Mayor and his graduate student Didier Queloz chose 142 Sun-like stars that they knew were single stars. Starting in April 2004, they spent a week of nights each month monitoring the stars for spectroscopic wobbles. It took only several weeks of watching to get the first inklings that something was up around the star 51 Pegasus. But by Christmas, the star disappeared from the observatory's view of the night sky, and the astronomers had to wait until July 1995 to double-check their results. After that, it took only a single week of observing to confirm their find, an exoplanet game winner that stunned astronomers: 51 Pegasi b, commonly called 51 Peg b, a Jupiter-sized planet so close to its star that it orbited in only four days.

"Mayor was one of the last people to enter the field," says Alan Boss, an exoplanet theorist who has closely followed the historical trajectory of the search. "But he came from a very different mindset. He came from the field of binary stars. So he was used to searching his data for even very short period companions, because binary stars can be so close that they're almost touching to so far apart that we're not sure if they're actually going around one another. So he was used to searching his data for all kinds of variable periods."

Gordon Walker—one of the scientific referees sought out by the editors of the scientific journal *Nature* to review Mayor and Queloz's historic paper before it was announced to the world—was among the first to receive the news. "Nobody, but nobody, suggested there were going to be Jupiters in few-day orbits," says Walker. "In looking for the familiar you miss the obvious." In fact, while Mayor and Queloz were waiting to confirm their finding, one of the world's leading exoplanet theorists published a paper in the journal *Science* arguing that Jupiter-sized planets would be found only at distances from their star that are similar to the distance Jupiter is from our Sun. Michel Mayor and Didier Queloz had cracked a millennia-old frontier and turned talk of probabilities and possibilities into the real realm of exoplanets.

DR. SEUSS'S UNIVERSE

In Jackson Hole in 2011, there was little time for looking back. The pioneer era in the search for alien worlds had turned into the hottest field in twenty-first-century astronomy. Whereas the pioneers found it difficult to get telescope time if they mentioned looking for distant worlds, exoplanet hunters today go to the front of the telescope line. In the second decade of the twenty-first century, exoplanet exploration has become big business: around the world there are now about a hundred ground- and space-based missions in search of exoplanets.

"Bruce Campbell is the father of all of the exoplanet work happening in the world right now," says University of California–Berkeley astronomer Geoff Marcy, one of the world's leading exoplanet hunters and the chair of the Extreme Solar Systems II conference. "There are several thousand people working on exoplanets, and it all goes back to Bruce Campbell." Marcy has codiscovered hundreds of the exoplanets found so far. He's done it using a technique pioneered by Campbell and Gordon Walker, who, he says, "invented the technique that we stole." In the cast of characters involved in the exoplanet search, Marcy plays the role of a central witness, having worked in the field since 1982. "If it wasn't for Bruce Campbell, you wouldn't be talking to me."

The Doppler technique is now one of a half-dozen proven ways to detect these once-unattainable alien worlds. Today, the search for exoplanets has the feel of the urban morning commute—focused, forward-looking, a little frantic. Many searchers worry they'll be just a little late to the big find. One bellwether of where the scientific world is at in its surveying of alien worlds is this: only fifteen years after the discovery of the first alien world, the discovery of a single, run-of-the-mill exoplanet will probably be lumped into an overview scientific paper rather than getting its own marquee billing. Just another alien world.

The irony in this age of exoplanet discovery is that the more we learn, the less certain we are. Twenty years ago, astronomers were more confident about the types of planets and how they form than they are today. After all, they had our Solar System as a shining exemplar. Stargazers had spent several hundred years figuring out Earth's Solar System siblings—where they were, how they behaved, and what they're made of. As a result, many astrophysicists thought they understood what a solar system is and thus had a template for all they'd find elsewhere in the cosmos. This is what they went looking for around other stars. Just as at home, any other solar system would start closest to its central star with smallish, rocky planets. Next, after a sizable safety zone of a gap inhabited by the asteroidal leavings of solar system formation, would come the giant gaseous planets, akin to Jupiter and Saturn. Still farther out are the medium-sized icy planets, the Neptune and Uranus lookalikes.

The planets orbit the Sun, all in near-circular orbits; not perfect, divine circles but only slightly squeezed ellipses. Gravitationally bound to one another, the whole system lies on a single, simple plane, flattened out like a spinning ballerina's skirt around the star's midline. We had our Solar System family photo, the one that adorned schoolrooms around the world, and that picture guided us in our search: every planet with a place, and every planet firmly in its place. It was a twentieth-century scientific version of fifteenth-century Ptolemaic order and beauty. Exoplanets changed all this.

If there's one thing astronomers have learned on their journey out, it's that our Solar System isn't the cosmic norm. "Almost every discovery in the domain of extrasolar planets was not expected," says Michel Mayor, who now leads the team that uses HARPS, the High Accuracy Radial Velocity Planet Searcher, a spectrograph on the 3.6-meter telescope at the European Southern Observatory's La Silla Observatory in Chile. HARPS holds the world record for parsing a star's speed down to just half a meter per second, the pace of a fast-crawling toddler. Rather than living in a cosmic-template solar system, we live in one amid an incredibly diverse cosmic zoo of exoplanets, more akin to a Dr. Seuss menagerie of colorful worlds than to a planetary artist's neat view of our home Solar System. This discovery has required the creation of a new language of exoplanets, from hot Jupiters, dwarf planets, and orphan planets to water worlds, carbon planets, and super-Earths. In a classic case of extrapolating from singular anecdotal evidence, we've had a severe case of solar system myopia. The psychological impact of our discovery of exoplanets is akin to a small-town girl arriving in a big cosmopolitan city and spending the first week rubbernecking at people sporting multiple piercings, speaking foreign languages, and displaying unimaginable behaviors. Our Solar System, for so long immense, all-encompassing, and defining, is now a mere village amid a multicultural galaxy.

At the Extreme Solar Systems II conference in Jackson Hole, much of the buzz concerned the fact that exoplanet hunters had entered a new era. No longer focused on singular planets, the larger debates are about understanding overall galactic rates of occurrence of different types of planets. With the number of putative exoplanets pushing several thousand, astronomers are drawing broad conclusions about what's "out there." They are getting the first broad-brush view of other worlds, and it is already clear that there aren't just the occasional exoplanets around other stars, but there are also exo–solar systems. When we look up into the night sky, we're seeing stars; but what our eyes can't see is that around many of those stars are swarms of orbiting planets. In the Stardust Revolution, the age of speculation is over: solar systems aren't an exception but rather a constant; in at least a quarter of the known instances, and probably more, stars and planets are born together.

These alien solar systems come in a smorgasbord of types. The first exo–solar system discovered was a trio of planets around the Sun-like star Upsilon Andromedae. The closest planet in the system, about two-thirds Jupiter's mass, zips around the star in just 4.6 days. Farther out, orbiting in about 241 days, is a planet with about twice Jupiter's mass, and orbiting its star once every three and a half years is the system's big brother, a planet of four Jupiter masses. The discovery of Upsilon Andromedae as not just a star but as a solar system awakened astronomers to a string of exo–solar system discoveries. The analysis of twenty years of Doppler measurements of the star 55 Cancri revealed an extended solar system with at least five planets and an architecture that looks more like a scaled-up version of our own Solar System, with four of the exoplanets closer to their star than the Earth is to the Sun. To date, our Solar System has the largest number of planets, but in time it will almost certainly lose this cosmic distinction. The Kepler-11 solar system, discovered in 2010, is a remarkable young solar system that includes at least six smallish, gassy planets, five of which form a tightly packed bunch that orbit between ten and forty-seven days. The alien solar system that looks most like ours so far is that around the star HD 10180, a Sun-like star in the southern constellation of Hydrus, or the Male Water Snake. Around this star, 127 light-years away, the HARPS team untangled the gravitational tugging of at least five Neptune-mass planets—between thirteen and twenty-five times the Earth's bulk—orbiting in as few as six days and as many as six hundred days.

Using space-based telescopes, it's now even possible to see alien solar systems in the throes of being born. Many people know the Hubble Space Telescope for its spectacular Hubble Deep Field image of early, and now very distant, galaxies. In 1994, after a major refurbishing mission, this eye-in-space caught glimpses just as stunning and revelatory of the process of cosmic creation—much closer to home. Hubble peered into the Orion Nebula, the closest area of intense star formation to Earth. There, amid the cold molecular clouds, were dozens of newborn stars, the stellar winds from them clearly carving out their natal clouds. Around dozens of these young stars was something no telescope had ever before resolved in such detail: protoplanetary disks. The Hubble images of individual stars were only thirty pixels square, the kind of blip that would make you think your computer monitor was failing. But the fuzzy blurs around the new stars let astronomers shorten the term "protoplanetary disk" to a cozier and more biological-sounding name, proplyds. Here weren't just new stars but new solar systems.

Later, when the Spitzer Space Telescope looked at stars in the infrared (seeing objects that Hubble couldn't), there again were stars with disks around them. The Spitzer's observations were informed by earlier infrared space-based telescopes that had measured infrared excesses coming from many stars—too much infrared light to be explained by the star itself. Something else, very near the star, was glowing in the infrared. On closer inspection of the data, the Spitzer astronomers realized they were looking at dust located in single, narrow rings around dozens of stars—warm dust glowing in the infrared. Protoplanetary disks are now thought of as an inevitable consequence of star formation, and they are orbiting almost all Sun-like stars. It's estimated that about ten solar systems are born in the Milky Way every year.

Differences notwithstanding, the language of our Solar System has been extended to describe the planets of other solar systems. Much like weight categories in boxing, exoplanets are grouped on the basis of their mass relative to that of the planets in our Solar System. In the heavyweight category are the Jupiter-like exoplanets—massive gas planets. Jupiter's mass is approximately 317 times that of Earth's, and exo-Jupiters range from about a quarter to about thirteen times Jupiter's bulk—at which point the giant ball of gas is a brown dwarf, a middling creature between planet and star. The midweight contenders are the Neptune-like exoplanets—worlds of about seventeen times Earth's mass. Farther down the planetary chain come the sought-after planets that weigh in at low multiples of Earth's weight.

The boxing analogy starts to fail when we consider that, for exoplanets, what's just as important, if not more important, than overall mass is what they're made of. For most exoplanets, this is extrapolated directly from their mass. Thus the giant planets are, like Jupiter and Saturn, thought to be primarily vast orbs of hydrogen and helium. The midsized planets, such as Neptune and Uranus, form from ices dominated by water, methane, and ammonia. The smallest planets are the ones dominated by rock and metal—planets such as Earth, Venus, Mars, and Mercury. For this solar system vernacular, exoplanet explorers have had to develop a new lexicon to describe the types of planets and planetary behaviors alien to our Solar System. Thus we have "hot Jupiters" (Jupiter-sized planets close to their stars); one astronomer has suggested classifying all such close-in planets as "roasters"; and there is also the possibility of a new class of waterworld planets—those whose entire surface is covered in liquid water.

It's still very much a developing and contested vernacular, in which the differences in language reflect competing views of what's out there—nowhere more so than when regarding the concept of super-Earths and mini-Neptunes. The discovery of a continuous range of planetary masses between those of Earth and Neptune—a class that doesn't exist in our Solar System—has been one of the great discoveries of the early exoplanet era. The big question is whether an exoplanet two and a half times Earth's mass should be dubbed a super-Earth or a mini-Neptune. The problem is that usually no one knows what the exoplanet is made of: Is it rocky, like Earth, and thus more conceivably habitable; or is it gassy, like Neptune? Some exoplanet astronomers believe that for exoplanets found to date, the term "super-Earth" is largely a sexed-up word used more to garner public attention than to accurately describe distant worlds that could just as well be, and probably are, Neptune-like.

However, we've yet to extend familiar-sounding names, like those of ancient Greek and Roman deities, to exoplanets. For now, they're named as stellar or telescopic accessories. Thus, the first exoplanet around an already cataloged star gets the star's name followed by a *b*. If the star is uncataloged, the exo-world gets the moniker of the discovering telescope—a number indicating the order of discovery by the telescope and a letter designating its order of discovery in an alien solar system; thus Kepler-10b is the name of the planet discovered by the Kepler Space Telescope.

On a more detailed level, what's emerging from the current exoplanet census taking is that the overall frequency of exoplanet types is quite different from that in our Solar System. It turns out that while our Solar System is dominated by mighty Jupiter's tug, planets this big are relatively rare. With the discovery of 51 Peg b, the first exoplanet, hot Jupiters appeared to be major players. The cosmic irony is that they've been like a sizzling movie trailer that belies the nature of the full-length movie. Based on the current exoplanet census, fewer than one in a hundred solar systems with Sun-like stars include red-hot giants. Similarly, only a tenth of all solar systems include a Jupiter-sized planet at all.

While Jupiter-sized planets appear to be relatively rare, it's the smaller planets that are ratcheting up the exoplanet tally. It may be that in the Milky Way overall, the number of planets increases with decreasing mass; in other words, there are far more small planets than big ones. One major survey of 166 Sun-like stars found that about 6 percent have a Neptune-sized planet. What's most striking is that the Milky Way is resplendent with an abundance of a planet weight class that is absent in our Solar System: the super-Earths, planets between three and ten times the Earth's mass. In our Solar System, we jump from Earth mass to Neptune mass, but the two largest exoplanet surveys to date—the European HARPS survey led by Michel Mayor and NASA's Kepler mission—indicate that between a third and a half of all Sun-like stars have super-Earth-sized planets with close-in orbits of less than fifty days.

Exoplanet scientists come in two basic flavors: those who like finding these alien worlds, and those who like thinking about how these worlds got to be the way they are. These are the astronomy equivalents of experimentalists and theorists. And when it comes to exoplanets, both groups in Jackson Hole were enthralled with what had been found out there. Exoplanets don't just come in a wild variety of sizes and orbital periods; they've also broken open theoretical thinking on how planets form and what planets can *do*. We've found not just different types of planets but also previously unimagined types of solar-system behaviors. Like in a family, it appears that solar systems emerge into stability only after a period of enormous sibling rivalry, great gravitational tussles in which young planets can be consumed by their stars or can be sling-shot out to spend billions of years wandering the cold, dark interstellar spaces of the Milky Way.

The biggest behavioral shocker for planetary theorists is the fact that giant Jupiter-like planets don't stay put where they're formed but often migrate enormous distances toward and away from their star. The logic goes like this: gas giants must form in the protoplanetary disk where there's lots of gas rather than dust, which would form rocky planets—that section of the disk is relatively far from the star, where Jupiter is in our Solar System. So what explains hot Jupiters? The thinking is that soon after birth, these giant planets spiral toward their star, drawn in by the gravitational and frictional dynamics of the depleting protoplanetary disk. Some of these migrating giants brake at orbital distances of only days; what causes this last-minute parking is still a mystery. Theorists now imagine that some doomed migrating planets do indeed continue on a death spiral, to be finally consumed and to disappear in the fiery maw of their star.

While some young planets get closer to their star, many others appear to be abandoned forever. In 2011, a joint Japan-New Zealand team released the results of a unique survey: in a search for exoplanets, they'd scanned the space not immediately around stars but in the vast reaches of interstellar space between them. The survey used a technique called gravitational lensing, which is based on Einstein's concept that gravity bends light. For an observer, *bend* is another way of saying *focus*. So the Japan-New Zealand team scanned a plethora of distant stars, watching for momentary intense brightenings—a gravitational lensing event when an object passes between the viewer and the star, gravitationally bending, or focusing, and thus causing the distant star to appear to momentarily shine more brightly. From the amount of brightening, astronomers can estimate the mass of the intervening object, whether it is another star or something else. The Japan-New Zealand team monitored about fifty million Milky Way stars every hour for almost two years. What it found came as a shock: ten free-floating, Jupiter-mass planets in interstellar space. So was born a new type of planetary category: the orphan planet. This initial sample led the astronomers to infer that there could be twice as many orphan planets as stars in the Milky Way. These dark bodies are orbiting like stars in the twirling stream of the galaxy. Where do all these lonesome planets come from? The astronomers in this study believe that the evidence points to planets that were gravitationally ejected from the stellar nest by close gravitational interactions with their star or other planets. Subsequently, other researchers have suggested that many solar systems end up adopting interstellar orphans as their own.

Given all these possibly eaten or orphaned planets, no single finding more impressed exoplanet theorists, experimentalists, and *Star Wars* fans than the discovery, announced in Jackson Hole, of Kepler-16b—the first circumbinary planet (a planet that orbits two stars). As occurs on Luke Skywalker's fictional home planet of Tatooine, on Kepler-16b, each day involves a double sunset. The cold, gaseous planet orbits a pair of stars, both smaller than the Sun, that in turn orbit each other. For planet hunters, this discovery opens yet another frontier. Until the discovery of Kepler-16b, astronomers didn't know if planets could form, or survive, in double-star systems. Kepler-16b isn't just a single new planet; it's also the first in a new class of planets that greatly broadens the possibilities for exoplanet exploration.

Rather than our own Solar System informing us about what we've found, the age of exoplanets has given us a new perspective on our home planetary system. Without doubt, the biggest local impact of exoplanets was Pluto's demotion to dwarf-planet status. When there were only nine planets in the entire cosmos, no one thought of ditching one, even if it was no bigger than distant asteroids. But a growing crowd of exoplanets, coupled with the discovery of other Solar System objects of equal or bigger size, resulted in Pluto becoming the first victim of planetary downsizing. Similarly, the realization that planets can migrate enormous distances from where they were formed, producing hot Jupiters, has inspired astronomers to take another look at Jupiter's history and possible past wanderings. It's now clear that solar systems are much more chaotic and interactive families than was thought before, with sibling gravitational interactions as planets grow from the protoplanetary disk having an enormous impact on any one planet's future.

The broader perspective provided by exoplanet behaviors has encouraged the view that Jupiter, as an infant, might have been like a sailboat caught in a gravitational current that took a grand tack from near its current distance from the Sun and moved to within a little more than Earth's distance, before being heaved back by Saturn's gravitational tug. The wanderings of this exoplanet giant would have set the rough outer limit for the formation of the rocky planets that appeared slightly later. An even more dramatic rewriting of our Solar System's early days is the possibility that somewhere out in the depths of interstellar space, Earth has a lost giant planetary sibling. One computational model of the early Solar System formation found that the current planetary alignments make sense only with the addition of an original fifth giant planet, one that was gravitationally ejected from the natal crib by a jumpy Jupiter.

But for many exoplanet searchers, the wonderful menagerie of known exoplanets is but the lead-up to the most sought-after type of exoplanet—a living one.

ALIEN EARTH

Amid all the excitement and anticipation over the extent and pace of exoplanet discovery, William (Bill) Borucki sat quietly on the podium beside two other panelists at Jackson Hole for an informal evening of exoplanet discussion. One front-row audience member stood up and, in a playful, congratulatory tone, said, "Bill, you always look so worried. You're already a hero. It's a revolution in planetary astronomy."

Both observations were true. As the persevering brainchild behind NASA's $650 million Kepler Space Telescope, the seventy-two-year-old Borucki, dressed modestly in beige slacks and a tan windbreaker, had turned the search for exoplanets from individual finds into a tsunami of discoveries. In just eighteen months since opening its electronic eye, Kepler had detected thousands of probable exoplanets and had shown that in time it will reap thousands more. But sitting there that evening, Borucki wasn't content. Many of the astronomers and astrophysicists searching for and studying exoplanets are primarily driven by the technical and scientific challenge; enamored with the motion of celestial bodies and solving complex multi-body motion problems; and building the complex mathematical, computational, and astronomical tools needed to do this. At this level, the exoplanet search is an extension of one branch of astrophysics that concerns itself primarily with celestial mechanics.

Borucki is as energized by the technical challenge as the next astrophysicist, intently discussing the nuances of the latest speaker's data during each coffee break. Yet that isn't what has driven him for the past three decades. He is part of an exoplanet researcher subclan for whom the search for exoplanets isn't ultimately about physics but about biology, about life. Borucki doesn't just want to find more planets; he wants to find the ultimate goal: another living planet, an alien Earth. The Kepler Space Telescope has detected thousands of exoplanets, but that's not why Borucki spent fifteen years doggedly proposing and developing the technology for the now-heralded mission at a time when, at best, he'd be dismissed, and, at worst, he'd be laughed off as a science-fiction-infected dreamer. Kepler's done a lot, but it has yet to achieve Borucki's lifelong goal of finding alien life or, in this case, a planet that ET might call home. Seated on the stage, Borucki knew that for all his decades of work, hope, and dreaming, it might not ever achieve this goal.

Borucki embodies the entire NASA vision and half-century history that encompasses both the dream of space travel and the way it has grown to merge with the search for extraterrestrial life and our astrophysical origins. In the mid-1950s, as a teenager in Delavan, Wisconsin—circus capital of the world and home to the original P. T. Barnum Circus—Borucki didn't find high school at all amusing. The school principal suggested a science club to keep him engaged, and the science teacher suggested projects—which Borucki promptly rejected in favor of his own: optical communication with UFOs. He led his 1950s crew-cut classmates in building a combination of ultraviolet, visual light, and infrared optical transmitters to signal any flying saucers passing over the US Midwest. They didn't make contact—didn't even complete the devices—but Borucki had found his passion: a way to turn the science fiction he loved reading into the tentative truths of science fact (his favorite is still Ray Bradbury's 1950 short-story collection *The Martian Chronicles*). In the process, he also learned a critical lesson that would guide the rest of his life: he might fail, but it was technically and scientifically possible to search for alien life.

If that was his first lesson, the next was on-the-job training in NASA's can-do culture. In 1962, the year after President John F. Kennedy committed Americans to go to the Moon within the decade, Borucki was one of the legions of young, talented engineers and scientists hired as part of NASA's Apollo program. The following year, his local draft board called him up for service in Vietnam, but, as Borucki puts it, NASA overrode them, saying his services were needed for another Cold War battle: to beat the Commies not in Vietnam but to the Moon. Fresh from college, Borucki was hired by NASA to help develop the Apollo reentry vehicle's heat shield. It was an awesome responsibility, at the very edge of engineering. Astronauts might make it to the Moon, but if the heat shield didn't have the right stuff, the crew would, like a burning meteor, be incinerated as the reentry vehicle plunged into the Earth's upper atmosphere. "It was like dipping the heat shield into the Sun for several minutes," says Borucki.

Borucki became part of a NASA team that worked around the clock, in shifts, seven days a week, to develop the heat-shield technology. The Apollo crew would reenter the Earth's atmosphere, traveling at a blistering six miles a second. To simulate the pressure and temperatures at this speed, Borucki and colleagues used two artillery cannons from a mothballed US Navy battleship. When the cannons were fired, the blast was so powerful it lifted the roof of the building in which they were housed. For all this brute force, Borucki's job focused on understanding the atomic minutiae at the interface between the atmosphere and the heat shield, where a superheated plasma formed—atoms shorn of electrons and emitting intense radiation. To do this, he built a state-of-the-art spectroscope, using lenses developed for US Air Force spy cameras, to study the light emitted by the plasma. Analyzing the nature of this light was critical to knowing the types of radiation and radiation flux the heat shield needed to endure. In the end, the team developed a graphite-epoxy-based shielding that brought successive Apollo astronauts safely home. Borucki learned about the enormous value of what could be learned from the study of light, and that with determination almost anything was possible.

When the Apollo program ended, Borucki began his own space research missions aimed at the search for alien life. First, he spent a decade extending his research in plasmas by exploring lightning on other Solar System planets. When an electrical charge passes through the atmosphere, it creates superheated plasmas, which Borucki and others thought might be the spark for forging prebiotic molecules, the building blocks of life. But in the early 1980s, when astronomers like Bruce Campbell were talking up the search for planets around distant stars, Borucki came across a quixotic scientific paper by Cornell University computer scientist Frank Rosenblatt that would change both Borucki's life and the search for alien worlds.

Rosenblatt's paper is one of the great, surprising gems of the Stardust Revolution. In the 1950s and '60s, Frank Rosenblatt was a computer scientist at Cornell University, where he developed the Perceptron, the first computer that could learn how to use a neural network that simulated human thought processes. But Rosenblatt was a prodigious polymath who combined his computer science research with interests in the psychology of perception—doing research in how air force pilots perceive depth and distance when landing on aircraft carriers—and astronomy, with a homemade observatory at his home outside Ithaca, New York. He talked about the stars and the search for alien life with leading Cornell astronomers, including Carl Sagan and Edwin Salpeter, who'd contributed to figuring out how stars make the elements.

While most Americans were focused on the Moon landings, Rosenblatt was already imagining exploring for more distant cosmic terrain. His 1971 paper, "A Two-Color Photometric Method for Detection of Extra-Solar Planetary Systems," proposed a two-pronged new way to search for these systems. First, Rosenblatt imagined that when a "dark companion" passes in front of its star, as seen from Earth, this transit, or mini-eclipse, would be visible as a tiny, transient dip in the star's light. In essence, the star would make the subtlest of winks. This stellar dimming could be detected by a telescope equipped with a photometer, a device for measuring the intensity of the light. Second, Rosenblatt noted that you didn't need an army of astronomers to keep watch for these transits; instead, he outlined a computerized system that could automatically watch stars and periodically record their brightness. It was the only astronomy paper Rosenblatt ever wrote. Six months after the paper was published, he drowned in a boating accident.

But Rosenblatt's grand planet-finding idea lived on in print, and with it, Bill Borucki had the project he'd been looking for all his life. If there's extraterrestrial life, it has to live somewhere. So the first step in trying to make contact is to find out if there are indeed other Earth-like planets and, if so, where they are. At the time, the other exoplanet-hunting techniques considered for NASA missions, such as stellar astrometry—observing a star's infinitesimal movements back and forth across the sky as a result of an orbiting planet—were sensitive enough to possibly detect Jupiter-sized planets. The beauty of Rosenblatt's transit technique is that it would be able to spot smaller planets, maybe even Earth-sized ones. Here was a way to actually search for alien Earths.

From his first paper on the topic in 1984, Borucki began to build a vision and an informal team of researchers with the goal of detecting Earth-sized planets orbiting other stars. As this vision developed, several things became clear. First, it would be more difficult than Rosenblatt imagined. To see small planets, they'd need to develop more sensitive light meters, or photometers. More importantly, Borucki and others calculated that the Earth's atmosphere was simply too turbulent and messy a lens to see through if they were hoping to spot minute changes in a faraway star's brightness. To spot the eighty-four-parts-per million fading of a distant star by a transiting exo-Earth, a telescope would need to be beyond Earth's atmosphere, in the darkness of outer space.

By 1992, Borucki's vision had morphed into a mission proposal, one that, like Apollo, would go big from the start. The goal wouldn't be just to find a single other Earth but to take a census of tens of thousands of Milky Way stars to determine the percentage of them orbited by Earth-sized planets. It was a broader scientific goal: not just are we alone, but also just how populated—at least with Earth-like planets—is the Milky Way?

To achieve this, Borucki's mission had to go beyond just spotting

Earth-sized planets to identifying those in their star's Goldilocks zone. Starting in the early 1960s, astrobiology pioneers began considering what a planet around another star would require to sustain life. As in all real-estate questions, at the top of the habitability list was location. A planet would need to be in its star's Goldilocks, or habitable, zone—a distance from the star within which the temperature would be not too cold and not too hot but would be just right for there to be liquid water and thus, potentially, life. The exact position of the habitable zone would be different for each star-planet system, dependent primarily on the star's size and the planet's atmosphere. A small drop in star size makes a sizable decrease in energy output. Thus, for smaller stars, the habitable zone is much closer to the star than is the case in our Solar System, where the zone is thought to stretch from this side of Venus's orbit to before that of Mars. In practice, however, each alien world's habitability would require individual examination, because a planet's atmosphere plays a crucial role—as the difference in habitability between Earth and the Moon quickly makes clear.

At NASA-Ames, Borucki now occupies the former office of long-time Kepler colleague and friend Kent Cullers, the outline of whose name is still visible as a palimpsest from the four-inch-high letters in black electrical tape he'd used to mark his locale. Cullers, the first astronomer to be blind from birth, turned not to light but to math and sound waves as a way of exploring the cosmos. He developed the first advanced mathematical codes to detect signals coming from alien civilizations, codes that became the core of the search for extraterrestrial intelligence (SETI) program. They were a perfect pair: Borucki wanted to find extraterrestrial intelligence; Cullers wanted to communicate with it. Together, the dreamer, the blind man, and other early members of the Kepler mission calculated the number of stars they thought they'd need to study in order to get a reliable Earth-like exoplanet census. In 1996, they suggested thirty-four thousand stars, but as Borucki continued to pitch the project and as technology improved, they broadened the search to one hundred and fifty thousand stars. If NASA was paying the same price, why not offer more stars? As Borucki put it, "Would you rather pay for something that does fifty thousand or one hundred and fifty thousand stars?"

While he developed the mission concept, he also led the proof-of-concept and R&D to demonstrate the underlying technologies. He spent years developing quantum-perfect charge-coupled device, or CCD, photometers, which were able to sense the energetic *ping* of a single incoming photon. At the Lick Observatory in California, Borucki and his team showed that automated photometry was possible by building Vulcan, a robotic system that monitored the light of ten thousand stars simultaneously.

Yet, for almost a decade, Borucki's exoplanet mission proposal repeatedly failed to make the grade at a NASA focused on the Space Shuttle, the International Space Station, and the Hubble Space Telescope. After all, why look for alien Earths if no one's even found a single planet of any size outside our Solar System? The discovery of 51 Peg b in 1995 changed all that. Featured on the cover of *Time* magazine and in scientific journals was the dawn of a new era: worlds around other Suns. Then, in 2000, the first exoplanet was discovered based on its transit of its star. Rosenblatt's technique worked. Finally, in December 2001, after four formal proposals, Borucki's mission, renamed Kepler, had wings. On the evening of March 6, 2009, Kepler lifted off from Cape Canaveral Air Force Station atop a Delta II rocket to begin its planet census mission. Borucki's dream of searching for alien worlds was now a $650 million space-based telescope, one of NASA's Great Observatories, and a key part of its Origins Program.

On the eve of the Extreme Solar Systems II conference, the baton had passed for the moment to Natalie Batalha, Kepler's acting science team leader. Batalha didn't sleep the night before the conference. As an astronomer, she's accustomed to long, sleepless nights. That night, however, was different. She wasn't operating a telescope, the usual reason for pulling an all-nighter. On that cool night in September, the first snow forecast at altitude in the Teton Mountains for the end of the week, Batalha wasn't imagining finding alien worlds; she was hurriedly crunching the numbers on the worlds Kepler had already discovered. If she stepped outside her motel-like room at the Jackson Lake Lodge and peered into the night sky, she'd see Kepler's field of view: just off the plane of the Milky Way, to the right of Cygnus, the Swan constellation, a swath of night sky—about the area covered by your hand held out at arm's length—including more than 150 thousand stars, ones that Kepler relentlessly monitored for evidence of transiting planets. It was Batalha's job to summarize the last three months of Kepler's planet-hunting data, the endpoint of an exhaustive process of weeding out potential false alarms.

Given the history of exoplanet false calls, the Kepler project treats possible planet detections in the same way that bartenders want to see two pieces of identification before they start pouring drinks. Whenever Kepler's data-analysis system senses three near-identical transits around a star, the team's software automatically produces a twenty-page Threshold Crossing Event (TCE) report. Then the humans, the TCE review team led by Batalha, get to work on scrutinizing the data and seeing if it truly fits a transit pattern as expected from an exoplanet, or if it is an aberrant light blip. They analyze the collection of photons detected by Kepler's photometer, or light meter, that form the stretched-out "U" of a transit—as if suddenly a star's straight line, its constant brightness, detoured sharply down and then back up. If it's a keeper, the TCE changes acronyms and becomes a KOI, a Kepler Object of Interest. Using this rigorous vetting process, the team estimates that nine out of ten planet candidates that make it to the KOI are indeed other worlds. Final confirmation that a KOI is an exoplanet—that there's indeed something there—depends on the KOI passing a series of further tests, including detecting the putative exoplanet via the radial velocity technique. For this, the Kepler team uses the Keck Telescope—at $90,000 a night, no refunds if it's cloudy—one of the few in the world with an accurate enough spectrometer to measure the tiny wobble induced by small exoplanets at the distance of those detected by Kepler.

Prior to the Wyoming meeting, Batalha had already gone over

each of the individual possible planetary detections with the Kepler team—putting aside some as too uncertain—and now it was time to compute the detections into a summary form. By six in the morning, Batalha had finished writing the computer code to process the latest Kepler data.

"I was sitting in my hotel room, looking out at the forest—it was a beautiful scene here in Jackson Hole, just savoring the moment before hitting the 'go' button to run my code to get the new numbers," she says. In the minutes it took the code to run, the cosmos got a lot more crowded, and alien worlds got a little closer. "It was a surreal moment to see that the number of Earth-sized planet candidates went up by 100 percent."

In fact, the next day—with no time for a press release, let alone a scientific paper—Batalha announced that Kepler had discovered another five hundred putative planets around distant stars, some of them appearing to be not much bigger than Earth. It was a historic, unprecedented announcement, almost doubling the number of known planets in the universe. On the same morning, Michel Mayor's HARPS team—Kepler's main competition in the race to find an alien Earth—announced a haul of fifty new exoplanets. This total included sixteen exoplanets that the HARPS team called "super-Earths," exoplanets they estimated weighed in at between one and ten times Earth's bulk, the smallest of them just three and a half times Earth's mass.

For Bill Borucki, the thousands of exoplanets that Kepler has discovered until now are stepping stones on the way to the ones he wants to find, the ones he wants to share with the world. Kepler is working its way down to smaller, Earth-sized planets—if they exist. "If the answer is no, there aren't [a large number of stars with such planets in their habitable zone], then there probably isn't much life out there, and it's going to be extremely difficult to find it and you might as well wait a hundred or a thousand years until the technology is much, much more capable," he says. "On the other hand, if you find lots of Earths, and particularly if a good fraction of them are in the habitable zone, then you can gather the enthusiasm to do the next steps . . . Kepler has to get such good data that we can convince the world to go out to explore the galaxy."

Less than two decades after the hard-won discovery of the first exoplanet, most astronomers would now bet their telescopes that a multitude of Earth-like exoplanets are there for the finding. This assessment is based on a solid planetary trail. It's clear that the Milky Way abounds not just in stars but in solar systems. The discovery of exoplanets has been a case of metaphorically finding the biggest, easiest-to-find ones first. Exoplanet hunters have burrowed down from big, Jupiter-sized worlds to mini-Neptunes and super-Earths. Astronomers are confident that the exoplanet size gradient doesn't hit a planet-building wall here—there's a consistently occupied neighborhood of exoplanet sizes, a gradient that gives every indication of continuing down to Earth-sized planets and smaller. It's a view that's corroborated by evidence from infrared observations of newborn stars whose protoplanetary disks, areas of planet formation, reveal the glowing, dusty raw materials for rocky planets. The most enthusiastic exoplanet theorists think the evidence points to the conclusion that every Sun-like star harbors in its light an Earth-like planet.

On a Monday morning in December 2011—for most Americans, a back-to-work day wedged between America's Black Friday shopping frenzy and the Christmas end zone—Bill Borucki and the Kepler team announced they'd landed their first Earth-like planet: a confirmed super-Earth orbiting in a Sun-like star's habitable zone. After watching three transits and verifying the planet's presence via ground-based telescopes, they announced Kepler-22b, a planet just under two and a half times Earth's radius orbiting a star six hundred light-years away. What made it the stuff of news headlines was that this distant world orbits its star, one slightly smaller than the Sun, in about 290 days, putting it, according to the Kepler astronomers, squarely in its star's habitable zone. Here was a planet that the Kepler team believed could be a large, rocky planet with a surface temperature of about 72°F, comparable to a comfortable spring day on Earth. Or, as the Kepler team knew well, it could be a mini-Neptune, a largely gas planet, perhaps one with a thick atmosphere that, so close to its sun, made it more similar to scorching and lifeless Venus than to a blooming spring day on Earth. It was the latest success in a three-year spree that's opening our eyes to a new cosmos, one resplendent with planets. In early 2012, based on these successes, NASA extended the Kepler mission to 2016. This additional time is essential for Kepler to spot smaller planets with long orbital periods, those that require numerous years of observation to record three separate transits, and for confirming Earth-sized planets around stars that have variable baseline brightness, which makes a transit detection more difficult.

Coming from the age of exoplanet speculation, we stand on the edge of the age of exploration. Thousands of exoplanets beckon. Somewhere out there in the night sky, circling another life-giving star, is in all likelihood another rocky world, its atmosphere the planet's churning breath. When we find it, we won't have found an alien world but a distant relation—a living planet, born of the same stuff, the same cosmic processes, chances, intense bondings, and violent breakings as our Solar System and the planet we call home. The discovery of that alien Earth will open the next chapter in the Stardust Revolution—one that, for the first time, we will be able to use to compare the story of planet Earth with that of another living planet; to see ourselves reflected in another's story.

CHAPTER 9

DARWIN AND THE COSMOS

We have had a century in which to assimilate the concept of organic evolution, but only recently have we begun to understand that this is only part, perhaps the culminating part, of cosmic evolution. We live in a historical universe, one in which stars and galaxies as well as living creatures are born, mature, grow old, and die.

—George Wald, Harvard University biochemist and Nobel laureate, "The Origins of Life," 1964

THE BIOLOGICAL BIG BANG

I've come to the Harvard-Smithsonian Center for Astrophysics—the CfA—in search of the current intersection of astronomy and biology, and I find myself literally standing at it. The center sits on a pleasant knoll in leafy downtown Cambridge, Massachusetts, at one end of Linnaean Street, named for the great eighteenth-century Swedish botanist and zoologist Carl Linnaeus. It was Linnaeus who developed the modern classification system for plants and animals based on shared physical characteristics—stems and leaves, fins and feathers. Linnaeus also developed the binomial species-naming system, giving us in 1758 the name *Homo sapiens* (meaning "wise man")—a mammalian species known for asking, in moments of self-reflection, such questions as "Who am I?" and "Where have I come from?"

Atop the knoll, a minute's walk from Linnaean Street, sits the Harvard Observatory. Built in the nineteenth century, it's tiny in comparison with the modern behemoths atop isolated mountaintops, but in historical perspective, this red-brick observatory is a giant. It was here, a century ago, that a scientific cadre of female astronomers, working for male professors, did the legwork in developing the modern classification and naming system not for species but for stars. The astronomers at this observatory photographed and analyzed the light fingerprints of tens of thousands of stars, the original images stored today in the observatory's basement vault. Just as Linnaeus grouped animals on the basis of shared characteristics, two hundred years later, Annie Jump Cannon and her colleagues grouped the stars on the basis of their common characteristics, the atomic lines that dominated each star's light fingerprint—here calcium, there iron. The result is the O, B, A, F, G, K, and M (*Oh Be A Fine Girl/Guy, Kiss Me!*) stellar classification system, adopted in 1922, that opened the door to a deeper understanding of the stars' origins and evolution, just as the Linnaean system pointed the way to questions about the origin of Earthly species.

Charles Darwin died in 1882, forty years before Cannon and her colleagues grouped the stars into their own families, but in the last sentence of *On the Origin of Species*, Darwin alludes to the Earth's cosmic context, to the way in which the evolution of our planet is tied to a larger story. He refers to the Earth as a planet that has long "gone cycling on." With this reference to our planet's yearly journey around the Sun, the father of evolutionary thinking leaves readers to contemplate the central importance of the Earth's astrophysical context to evolution. In his parting words for those thinking deeply about evolution, Darwin pointed the way to the cosmos. Today in the Stardust Revolution, our story—the epic journey of our origins, the tracing of our family tree—doesn't start with thinking about the origin of species but rather with thinking about the origin of stars. To understand the cosmic context of our origins, of cosmic evolution, is to trace the genealogy of stars back to the dawn of time. In the Stardust Revolution, the long-asked cosmic question "Are we alone?" has been transformed into the extreme genealogy question

"How are we connected?" Perhaps in the not-too-distant future, the question may be "How are we related?"

In search of the very base of our cosmic family tree, I've come to the CfA—which, in the words of its website, is "the world's largest and most diverse center for the study of the Universe"—to meet the distinguished astrophysicist Alex Dalgarno. He's e-mailed me to say that he's in office B324. I walk up to the third floor, make my way to the "B" hall, and follow the numbered door labels into a short side hall, which includes the photocopy room. Here's room B323 and then B325. I look again, and my initial disorientation is confirmed. I get a sense that, in something akin to *The Hitchhiker's Guide to the Galaxy*, all the other rooms are there, but there *is no* room B324. It's a cosmic joke. I ask a graduate student about this alphanumeric mystery. He explains that B323, a locked door, is the entrance to an anteroom that leads to B324. B324 is there, but you have to know how to get there. It seems a fitting introduction to Dalgarno, since I've come to ask about the very beginnings of cosmic evolution, a time metaphorically hidden behind a series of doors for which astrophysicists have discovered some of the keys—none more than Dalgarno himself.

When I finally sit down with Dalgarno, the Phillips Professor of Astronomy, I find myself with an extremely genial eighty-two-year-old with prominent fly-away, bright-white eyebrows. He is slouched slightly forward in a wooden chair, its arms worn smooth with use. He wears a herringbone jacket that he keeps in his office and sensible black loafers. The spark in his eyes is a reminder of an earlier time, when Dalgarno, had he put more effort into soccer than into mathematics, might have had a career with England's Tottenham Hotspurs, with whom he was invited to try out. Few outsiders now come looking for Dalgarno, and those who work here know where to find him; he's been a fixture at Harvard since 1967. Along with this impressive tenure, he's had the ultimate insider's view on the development of late twentieth-century astrophysics as editor for three decades of one of the field's premier scientific journals, the *Astrophysical Journal Letters*.

As such, Dalgarno is both professionally and personally a bridge to earlier times. Born in 1928, he recalls listening to astrophysicist Fred Hoyle's famed BBC broadcasts in the late 1940s and, during his first academic posting in the late 1950s, seeing Hoyle lecture in Dublin, Ireland, where he won over a rowdy group of students with his intense presence and broad speculation on everything from the origin of the elements to the origin of everything. "Of course, half of it wasn't true," Dalgarno says with a chuckle. However, the half that was true, the part that fueled the astronomer's periodic table, has guided Dalgarno's career. Dalgarno has always loved tackling puzzles of all kinds, and he was drawn to the mother of them all. He's one of the founders of the field of molecular astrophysics, or cosmic molecular evolution—the study of how the universe has evolved its present chemical complexity. In simple terms, if you feel that the universe is getting more complex all the time, Alex Dalgarno has proven you right.

When it comes to the cosmos' chemical evolution, Dalgarno has probably spent as much time as anyone in the history of the planet thinking about the universe's most basic element: hydrogen. He's studied hydrogen as an ion and as a molecule; he's studied its isotopes, its quantum states, and its optical properties. The amazing thing is that he's considered hydrogen not so much to understand this tiniest atom but to understand the essence of the entire cosmos. The doorway to understanding the full complexity and expanse of the observable universe is through its simplest atomic constituent—which makes Dalgarno perfectly suited to think about the very beginnings of time, when *hydrogen* was just about the only elemental word the newborn universe could utter. "In the beginning," he says, "there were no atoms or molecules. . . . It was an unpromising scenario for the formation of complex structures like galaxies; black holes; stars; planets; nuclear, atomic, and molecular systems; and living organisms." Just as looking at newborn baby pictures often leaves us amazed to think that *that* little, reddish, prune-faced thing will grow up to be a suit-wearing banker pulling in a six-figure income, so, too, does the newborn universe appear almost unrecognizably different from what it will become. Perhaps the most revolutionary insight of

▲ "No, it can't be there," leading astronomers told Nobel laureate Charles Townes, who used a radio telescope to discover cosmic water in 1969. *Photo by the author.*

In the past forty years, more than 140 (mostly carbon-based) molecules have been discovered in space, including all of life's precursor molecules. *Image courtesy of ESA, HEXOS, and the HIFI consortium.* ▼

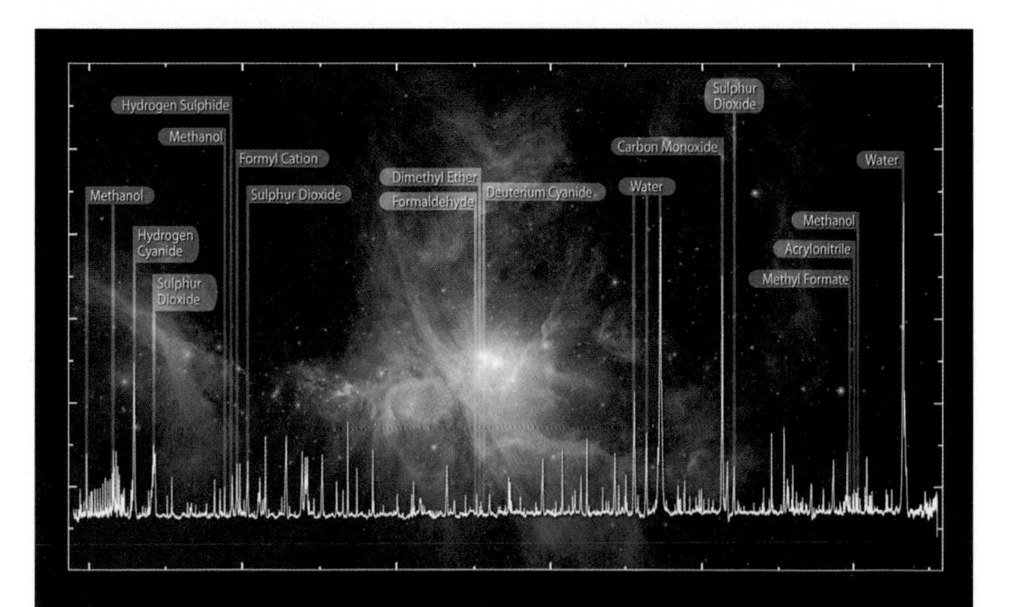

HIFI Spectrum of Water and Organics in the Orion Nebula © ESA, HEXOS and the HIFI consortium E. Bergin

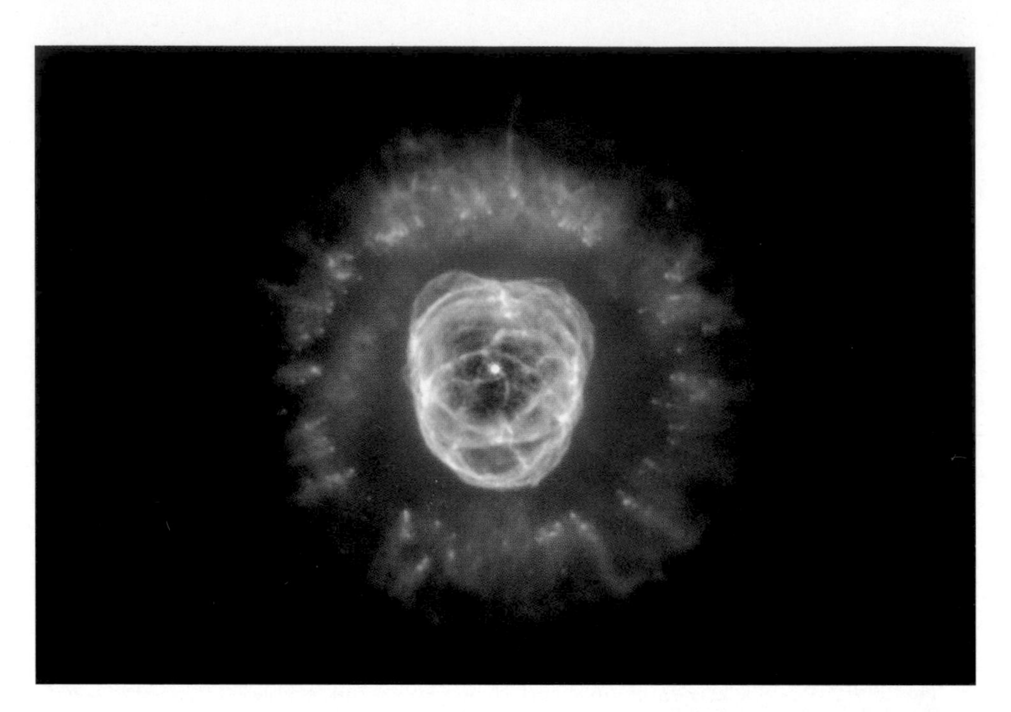

▲ At death, the bulk of a Sun-like star's body is transformed into dust and organic molecules, the building blocks of living planets, as in this image of the Eskimo planetary nebula. *Image by NASA, ESA, Andrew Fruchter (STScl), and the ERO team (STScl + ST-ECF).*

Supernovas, as seen in this composite infrared and x-ray image of the Tycho supernova remnant, seed the cosmos with vast amounts of planet-making dust and most of the elements in the periodic table heavier than iron. *Image by MPIA/NASA/Calar Alto Observatory.* ▼

▲▼ NASA-Ames astrobiologist Scott Sandford is realizing astrochemist Carl Sagan's dream of understanding life in a cosmic context, including showing in a NASA-Ames lab that amino acids can form in the cold cosmic clouds from which solar systems emerge. *Top and bottom images by the author.*

▲ The charcoal-like Tagish Lake meteorite, collected while still frozen in 2000, is a carbonaceous chondrite made up by weight of 6 percent organic material formed in space. *Photo courtesy of Michael Holly, Creative Services, University of Alberta.*

True stardust. An atomic-resolution transmission electron-microscope image of a pre-solar nano-diamond, about two thousand atoms in size, isolated from the Allende meteorite. *Image courtesy of Tyrone L. Daulton, Washington University, St. Louis, Missouri.* ▼

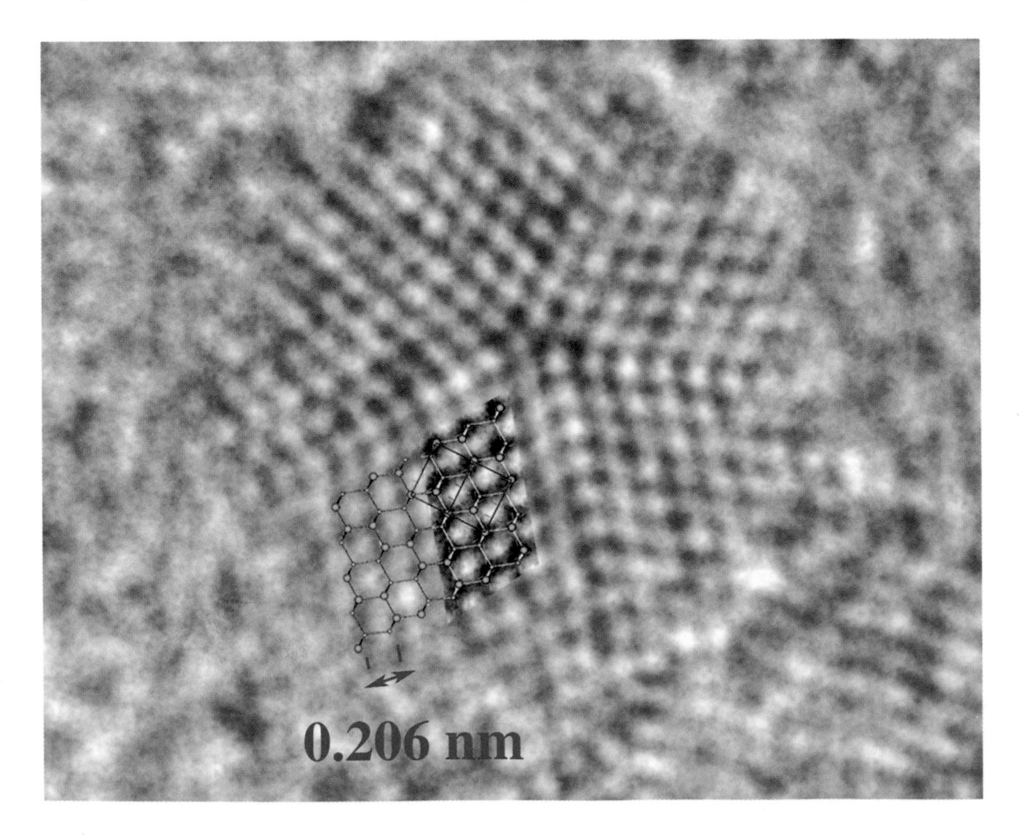

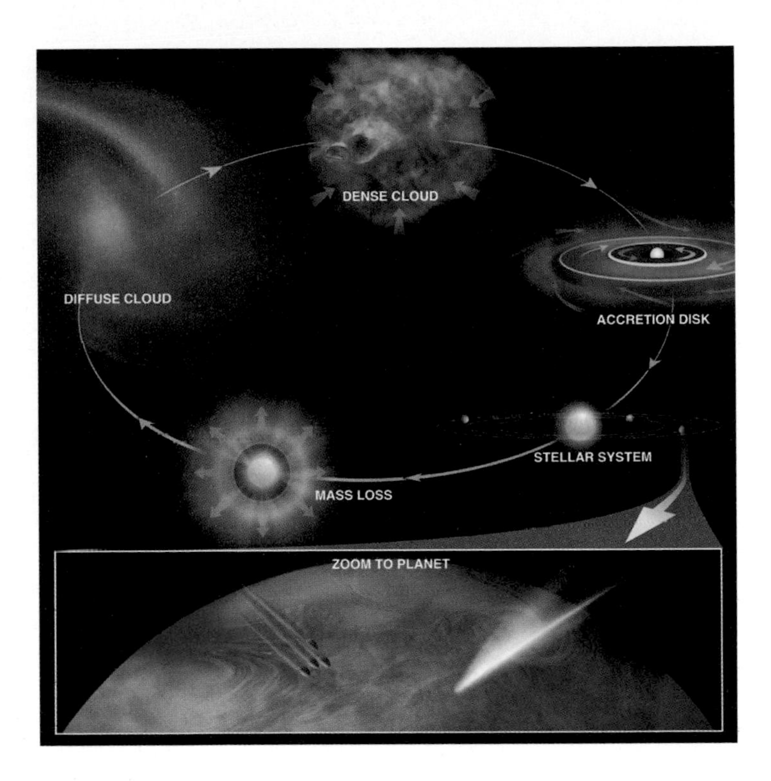

▲ The cosmic chemistry cycle includes the raining down of vast amounts of organic material to the surface of newborn planets. *Image by Bill Saxton, NRAO/AUI/NSF.*

Modern Galileos at the Extreme Solar Systems II conference in September 2011 (*from left*): Alexander Wolszczan, discoverer of the first planet-like bodies orbiting a pulsar; Michel Mayor, codiscoverer of the first exoplanet around a Sun-like star; Natalie Batalha, acting science team leader, Kepler Space Telescope; Bill Borucki, Kepler Space Telescope mission principle investigator; David Charbonneau, first to detect a transiting exoplanet; and Geoff Marcy, the most prolific exoplanet discoverer in history. *Photo by the author.* ▼

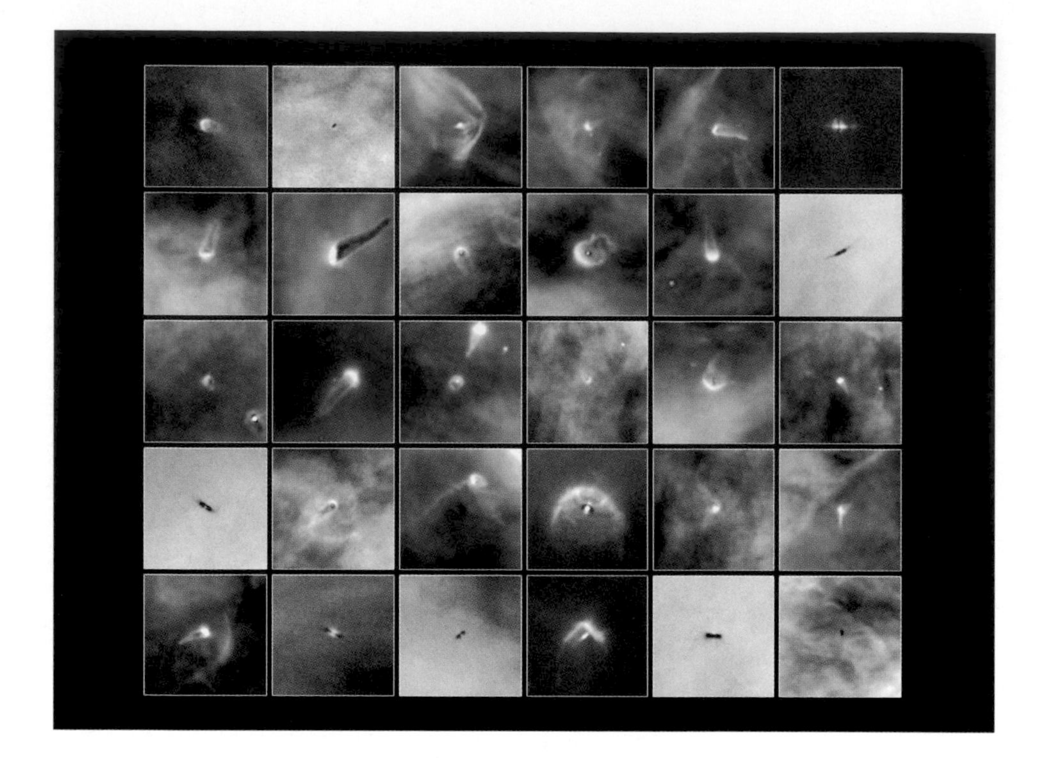

▲ The birth of new worlds. Spectacular Hubble Space Telescope images reveal the formation of embryonic solar systems, or proplyds, in the Orion Nebula. *Image by NASA, ESA, and L. Ricci (ESO).*

The Kepler-10 star system, located 560 light-years away, includes a sizzling rocky planet (the dark spot transiting the yellow sun) with a radius just 1.4 times that of Earth's, but so close to its star that it orbits in less than a day. *Image by NASA/Ames/JPL-Caltech/T. Pyle.* ▼

▲▼ Harvard University astrophysicist Alexander Dalgarno's research describing the origins of cosmic chemical complexity has helped lay the groundwork for others to consider the biological big bang and the nature of universal life. *Top image by the author. Bottom image reprinted courtesy of Steven A. Brenner*, Life, the Universe, and the Scientific Method *(FfAME Press, 2009), p. 312*.

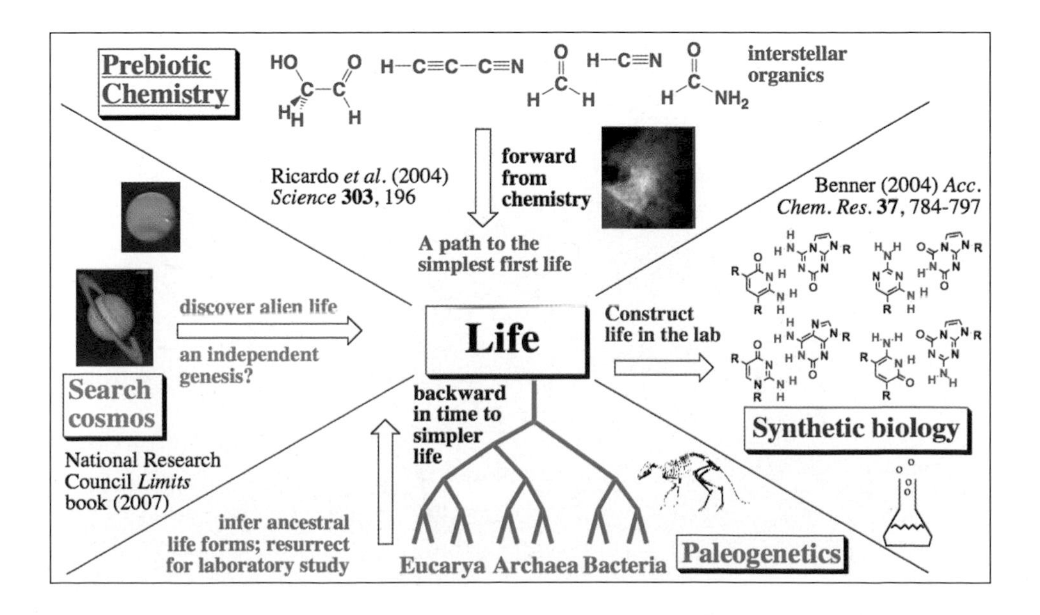

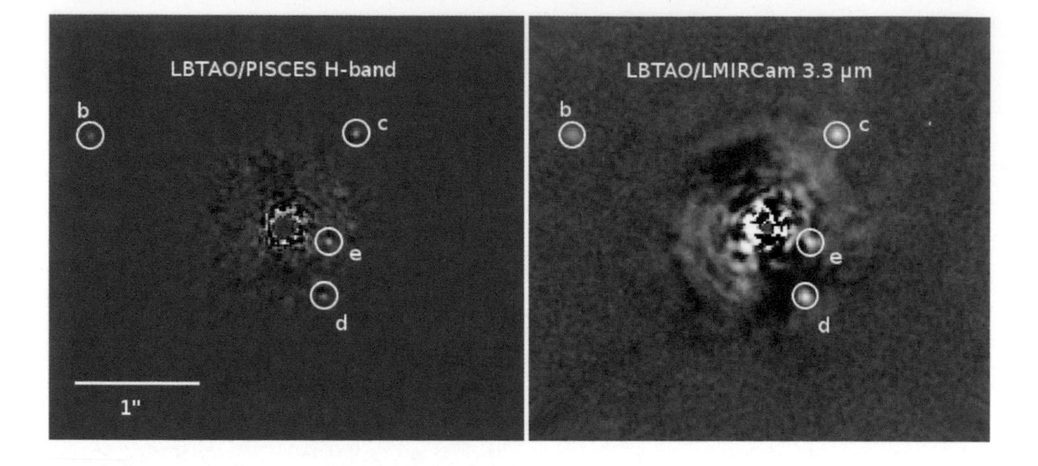

▲ A historic moment. We now have more direct images of exoplanets than of planets in our Solar System, including this image of four giant planets around the star HR8799. *Image reproduced with permission from the AAS.*

Earthshine, light from the Earth reflected back from the dark part of a crescent Moon, has been used to identify Earth's biosignatures, including the presence of vegetation and oceans. A similar technique would be used for taking the pulse of Earth-like exoplanets. *Image by ESO/L. Calçada.* ▼

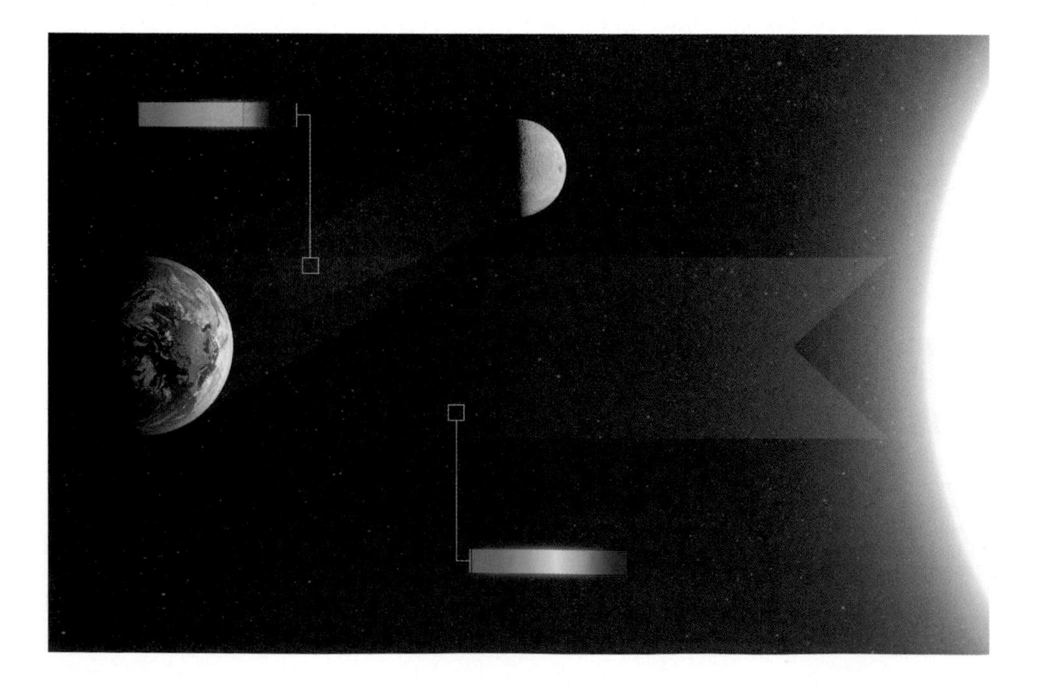

the astronomer's periodic table is that, for approximately the first two hundred million years in the infant universe, across all its vastness, there wasn't a single atom heavier than beryllium, not a single atom of carbon, nowhere the glitter of gold.

The short version of big-bang nucleosynthesis—the formation of the first atoms—might sound familiar to many couples: about fifteen minutes of intense procreative activity followed by a two-hundred-million-year lull in atom-making, most of it spent in darkness. At its first breath, the cosmos was unrecognizable to everyone except particle physicists. The universe emerged with such energetic fury that only the fundamentally indivisible particles, the quarks and electrons, could survive intact, bathed in a sea of intense radiation. The newborn cosmos was dense, gassy, and unimaginably hot, too hot for matter to stick together; there was nothing solid or liquid. Particles collided with such force that they ricocheted rather than bonded or fused. But as the universe expanded, creating time and space, it also cooled. After just one second, slightly less energetic quarks congealed into the more familiar neutrons and protons that form the nuclei of atoms. Within three minutes (astrophysicists clock it to about two hundred seconds, less time than it takes most of us to shower), the growing cosmos' temperature had fallen to under two billion degrees Fahrenheit. At this temperature, the cosmos experienced its first burst of nuclear building, or big-bang nucleosynthesis. Though most protons stayed single, a quarter by mass of all the baryonic matter (the stuff we see as matter, excluding dark matter) was forged into helium nuclei—two protons and two neutrons—and tiny amounts of lithium and beryllium, the next two elements on the periodic table. After fifteen minutes, this first act of heated creation was over. But the growing universe was still so hot, and thus energetic, that these nuclei couldn't hold onto a single electron; they were fully ionized. For about four hundred thousand years, says Dalgarno, the universe "coasted along" in this ionized state, expanding and cooling.

Then, like a baby's first smile, something truly new happened in the life of the cosmos: the first fully formed atoms were born. The expanding cosmos had cooled enough that, one by one, across space and time, when a positive nuclei and an electron collided, they didn't deflect like colliding billiard balls but stayed bonded together to form the neutral atoms of hydrogen and helium. Astrophysicists call it the Recombination Era; though the term *recombination* is a misnomer, since these negative and positive particles were never previously combined. With neutral atoms came a crucial evolutionary stage. The infant cosmos took its first chemical baby steps. As Dalgarno describes it, through a series of collisions between helium and hydrogen atoms energized by bountiful electrons and photons, most lone, neutral hydrogen atoms joined to become molecular hydrogen. These were the cosmos' first atomic bonds, and with them came the introduction of a new cosmic form: the molecule. With this step, the cosmos gradually went dark. Molecular hydrogen is a gassy curtain, opaque to visible wavelengths of light. To the human eye, the universe was completely dark for its first hundred million years—hence the Recombination Era's other name, the Dark Ages.

Where others see darkness, however, Dalgarno sees something else: the dawn of chemistry. Molecular hydrogen paved the way for the creation of a vastly expanded cosmic alphabet, one that could articulate the language of life. "The introduction of the neutral hydrogen molecule was a crucial step in the evolution of the universe," he says. It was only out of darkness that light could emerge. It was only because the cosmos went dark that the next evolutionary step could take place. Star formation depends on atoms' and molecules' abilities to radiate away the heat from gravitational collapse, and molecular hydrogen was the antifreeze of the early universe. Without it, the first stars couldn't be born almost two hundred million years later, in the era of cosmic dawn, now one of the most sought-after epochs in cosmic history. With the birth of the stars came an exponential bump in the nature of cosmic chemistry, the forging of a panoply of elements from carbon to uranium. These elements, in turn, found partners to create previously unseen types of unions, and they filled the universe with a new vocabulary of molecules such as carbon monoxide and formaldehyde—the building blocks of organic chemistry—and minerals, including sandy silicon dioxide, the crystalline beginning of rocky planets.

"I'm afraid I haven't been of much help," Dalgarno says with a smile and a shrug as I turn off my microphone and prepare to leave his office. He's referring to my quest to find the intellectual intersection of biology and astronomy. He's an astrophysicist; he doesn't feel comfortable leaving the realm of firm, testable numbers and equations. Linnaean Street might be a baseball's throw away, but biology is a foreign land, a foreign language. Yet Dalgarno's insights are central to the Stardust Revolution. With a physicist's penchant for exacting detail, he helped open the door to the conceptual framework that the cosmos isn't just expanding; it's growing more chemically complex in a process of cosmic evolution. Dalgarno is typical of many of the key protagonists in the Stardust Revolution: each has contributed a detailed piece—some small, some large—of a grand puzzle, often with little or no initial inkling of the larger picture to which they were contributing.

In documenting the emergence of cosmic chemical complexity, of a cosmos replete with large carbon-based molecules emerging from a simple soup of hydrogen and helium, Dalgarno and others have shifted the notion of the big bang from one relating to the origins of the physical universe to the story of a biological big bang. The tale he recounted to me in his office is usually heard as the crowning achievement of twentieth-century cosmology and astrophysics, joining the birth of the visible universe, of time and space, from the quantum nature of atoms to the grand architecture of the cosmos. The story Dalgarno tells is, however, a different one; it is as if a second narrator takes the stage in a Shakespearean drama and provides the audience with another view of the very same events. Dalgarno's version leads not to the nature of galactic structure nor to the dynamics of pulsars but to something much more deeply mysterious: life. In documenting the emergence of molecular complexity, he has mapped the thin, tendril-like roots of our cosmic family tree. He has helped join Darwin and the cosmos.

In the halls of the CfA today, Linnaeus has been given a central place. His term *homo*, for man, is derived from an ancient proto-Indo-European word meaning "earth" or "ground." Now the wise man of the Earth is wondering what shared characteristics tie him to the stars. At the CfA, astrophysics and biology have been joined in its Origins of Life Initiative, led by Dalgarno's CfA colleague Dimitar Sasselov, an astronomer turned astrobiologist. For Sasselov and his network of astronomers, physicists, chemists, and biologists, the search for an alien Earth—a living planet around a distant star—and synthetic biology—the ultimate goal of which is to make life in the lab from scratch—are intrinsically connected as part of the same question. The program's goal, according to Sasselov, is "a revolution in our understanding of life, and its place in the cosmos."

As Harvard's Origins of Life Initiative exemplifies, the Stardust Revolution is about much more than finding a single other instance of alien life. It's about developing a universal understanding of life and its origins and cosmic distribution. It's about understanding how Darwinian evolution can be extended, or embedded, in the breadth of cosmic evolution extending from the origin of the elements in stars, to cosmic chemical evolution, to the emergence of solar systems, and—of most interest to us—to the origins of life in this cosmic context. When we seek our origins, we follow the tree of life that connects us not just back to distant ancestors on Earth but also to everything else in the cosmos, allowing us to develop an understanding of life as an emergent weave from the atoms forged in stars to the birth of a child. At the heart of this pursuit, the tracing of our family tree back to a time before the first life on Earth or even before the Earth itself, is a truly extreme genealogical question: What *is* life?

WHAT IS "LIFE"?

The most interesting insights at science conferences, in my experience, occur during coffee-break kibitzing. The biennial NASA Astrobiology

Science Conference, held in League City, Texas, in April 2010, was a stellar example of this. During the conference—held just down the road from NASA's Johnson Space Center, where ground controllers guided the Apollo missions to the Moon and back—numerous presentations profiled the latest research results on the cosmic origins of life and the ways in which alternately some terrestrial life was later obliterated by cosmic collisions. What I didn't know then was that many of the learned astronomers, biologists, chemists, and geologists were avoiding one question in the way an attentive driver avoids a pothole: What, exactly, *is* life?

I encountered this intellectual road hazard only during a coffee break, when I found myself standing beside a friendly thirtysomething NASA engineer. We shared tidbits about home and family, and then I asked him what he did. He replied that he was developing science-on-a-chip tools that could chemically identify the presence of life for future NASA robotic missions to Mars and possibly to other Solar System bodies. I asked him what his sensor would search for, what he thought life was. He smiled broadly and replied, "Life is love." At first I thought he was avoiding the question or perhaps that he was being philosophical or folksy. He was doing a little of each, but he was also being completely honest. Although his answer wasn't the stuff of scientific journal articles, "Life is love" is in some ways as good an answer as anyone can provide. If love is the force of attraction between two beings or objects, then we view a cosmos as being defined by love from its first moments—quarks rushing to one another, hydrogen atoms bonding, helium atoms fusing, stardust joining to form planetesimals; all this great coming together.

In the Stardust Revolution, our understanding of life is getting a cosmic makeover. As astrobiologists consider the detailed origins of life on Earth and search for it elsewhere in the cosmos, the deeper problem they face is not how life starts but what life *is*. How will you know that you've found what you're looking for if you don't know exactly what you're looking for? What's so seemingly obvious—whether something is either alive or dead, an observation reminiscent of the great Monty Python "dead parrot" skit—is in fact the greatest conundrum to astrobiologists. This in itself is interesting. The weird thing about life, about biology, is that it's based on chemistry and physics. Physicists look for as-yet-undetected particles, but they're not wondering, "What *is* a particle?" There's matter and energy, and they just *are*. And chemists deal with the stuff of physicists, connecting atoms together to form molecules.

Yet for biologists, there's a big, usually unspoken of hole in the progression of things. At some point in the cosmic story of matter, magic seems to enter the equation in the form of life. How else can you explain going from stuff to life? Filling in this gap in our cosmic story is the grand challenge of twenty-first-century science. It's one that's very personal, when we consider that *we* are life, and thus the question could just as well be asked, "What am I?" The Stardust Revolution is a quest that is no longer the domain of only, or even primarily, biologists—of those who study what used to form the boundaries of "life"—it also involves chemists, physicists, astronomers, and mineralogists, among others. To understand the fundamental nature of life, these scientists are turning to molecular fossils and synthetic cells—and they're looking to the stars.

Fortunately, Steve Benner was also at the Astrobiology Science Conference in Houston. Benner thinks about, experiments on, and argues about the universal nature of life in the way that other guys get wrapped up in the details of old Ford® pickup trucks—which, if you've never talked with old-truck aficionados, is obsessively and in great detail. Sporting a frothy mop of blondish hair and radiating energy, Benner has a youthful pluck combined with serious smarts and curiosity that make him a natural intellectual scrapper. He's been brought to this conference to debate the origins of life and does so with humor and insight, even as his co-debater tosses audible backstage jibes during Benner's presentation. It's a disposition that's also given Benner, a chemist, the gumption to continually step outside the bounds of what most others think are the limits of the field.

Working in Harvard University's chemistry department in 1984,

Benner veered from his planned research to complete the then-outlandish idea of the first chemical synthesis of a gene that encoded an enzyme—a protein that catalyzes other reactions. In other words, his lab group built a string of DNA that was the chemical blueprint for an enzyme. In so doing, Benner helped to found the now-booming field of synthetic biology, a field that looks at building life the same way a garage mechanic rebuilds an engine from its constituent parts. The difference, of course, is that synthetic biologists are trying to build "engines" that in turn make baby engines by themselves—baby engines that, over generations, also evolve. Benner developed ways to resurrect ancient proteins from extinct organisms, a process once thought of as a branch of science fiction but that is now the fruitful field of paleogenetics. According to Benner, despite all this innovation, he was threatened with the loss of his US government funding from the National Institutes of Health for not focusing on his proposed research agenda. As a result, he accepted a no-holds-barred position at Zurich's Swiss Federal Institute of Technology, one of the world's top research universities and former home to Albert Einstein, as well as to twenty-one other Nobel laureates.

In his work at Harvard, Benner had become intrigued with how molecular aspects were connected to whole-organism evolution—with the evolutionary links between molecules and life. At that time, he says, "I was the only chemist in the world interested in evolution." Once again, however, those paying the bills thought Benner was outside the bounds of his discipline. Evolution was for biologists, said colleagues in Zurich, who told him that combining evolution and chemistry was *nicht salbre*, or intellectually dirty. But the time in Zurich gave Benner the space to develop the framework that now guides his research at his own institute in Gainesville, Florida, the Foundation for Applied Molecular Evolution, whose acronym—FfAME—perhaps speaks to Benner's larger ambitions. At FfAME, Benner is after nothing less than an understanding of life not just as we now know it but also as a universal phenomenon. About 170 miles northeast of Cape Canaveral, Benner is working to bring our understanding of life's fundamental nature into the space age. To do this, he and the dozen and a half other FfAME researchers are taking an integrated, four-part approach to understanding life as a universal concept, one that represents the spectrum of ways other stardust scientists are tackling the question.

Benner is looking both backward and forward at life by trying to make it and by guiding the search for it beyond Earth. Researchers at FfAME are using molecular paleontology to work backward down the tree of terrestrial life in an attempt to identify the molecular missing links to the origins of life on our planet. They're doing modern-day Miller-Urey-type experiments to explore how the rain of cosmic molecules on the primordial Earth may have given rise to the simplest first life. They're trying to construct life in the lab from scratch via synthetic biology—they're not trying to re-create how life actually emerged on Earth but rather are trying to see if they can create life de novo. It's this "build it and you will understand it" approach that many biochemists believe is the key to unraveling life's chemical nature.

The final slice of FfAME's research pie is devoted to considering how the question of what life is applies to the search for other life in the cosmos. It's this piece of his research that landed Benner the position of cochair of a US National Research Council committee called the Committee on the Limits of Organic Life in Planetary Systems, a subcommittee of the Committee on the Origins and Evolution of Life. The subcommittee's task, as requested by NASA, was to provide guidance for its scientists and engineers in the design of a new generation of life-detection experiments. In 2007, the committee, a veritable who's who of distinguished US astrobiologists, astronomers, microbiologists, chemists, and planetary scientists, released its report, *The Limits of Organic Life in Planetary Systems*.

Benner, who cowrote the report's final draft, knew then that it wouldn't include one key point: a definition of life. "This is no accident," he wrote, reflecting on the experience. "Early in the committee's deliberations a conscious decision was made *not* to include a definition of 'life' in the book. Perhaps this reflected cowardice. It may, however, be better viewed as an expedient based on wise experience. Nearly every member of the panel had spent hours in other committee meetings discussing that definition, with little productive outcome." Fortunately, Benner has a good sense of humor, which undoubtedly helps him survive the slippery slopes of defining life. The committee report is dedicated to "non-human life forms, wherever they are."

As we sit poolside under the hot Houston afternoon Sun, Benner walks me through the history of why, in the Stardust Revolution, the definition of life has become such a nebulous topic. It wasn't always so. The problem is that the definition of life has evolved. "Life" used to be so simple. Indeed, from the eighteenth century to the first half of the twentieth century, biologists didn't worry about defining life. Life was like a beating heart: you just had to look and listen, and you could tell something was alive. Ironically, all this began to change with what was at first seen as the answer to a definition of life, the discovery of DNA's structure. On February 28, 1953, Cambridge University molecular biologists Francis Crick and James Watson left the university's Cavendish labs and headed for the Eagle pub to celebrate. There, Crick is reported to have raised his glass and announced to his pub mates, "We have discovered the secret of life!" They toasted their victory in the heated race to discover the structure of DNA; they were the first to grasp its now-iconic double-helix structure, like two spiral staircases embracing one another step-by-step.

But Crick had overstated their case, though many consider his and Watson's discovery the greatest in twentieth-century biology. The structure of DNA didn't answer the question of the mystery of life; it greatly deepened it. Watson and Crick's discovery paper was published the same month that Urey and Miller published their paper about life's possible origins on Earth from a primordial soup of simple molecules. The combination of a new vision of the molecular basis of life in DNA and thinking about life's molecular origins soon muddied the waters of life's nature and led to a molecular-level chicken-and-egg problem. DNA is often referred to as the blueprint for life because the information it encodes directs the assembling of proteins; however, biochemists soon also realized that DNA itself is assembled by proteins. So the discovery of this molecule of life encoded a conundrum: Which came first, DNA or the proteins that assembled it? Similarly, Benner explains, sitting on a plastic chaise longue, the discovery of DNA also launched what he calls "the great moose debate." Prior to the 1960s, biologists were still largely tied to a Linnaean vision of life, that of the whole plant or animal—in Benner's example, the great antlered moose. However, the rise of molecular biology created a great divide, he says, between "those who studied moose and those who studied blended moose." In essence, the discovery of DNA pushed the quest to understand life down to the molecular level—in other words, not a whole moose, or even a mix of its cells, but the mix of chemical juice that makes a moose; that is, chemistry. This was anathema to many biologists, who felt their subject matter was metamorphosing into something unrecognizable, a point Benner drives home by saying that biology has become a subdiscipline of chemistry.

The extent to which life has become chemistry is denoted in NASA's 1994 working definition of life, a definition adapted from work by Carl Sagan and used by Benner as his guidepost: "Life is a self-sustaining chemical system capable of Darwinian evolution." There's nothing in this definition that rings true to what many immediately associate with life, such as movement or intelligence. Instead, in developing a prototype cosmic definition of life, astrobiologists have defined life in the broadest possible terms as something that can keep itself alive and have kids who in turn have kids who'll be different from their parents because of a combination of genetic mutations and population-level adaptation to changing environmental conditions.

Today, scientists generally agree that life on Earth began before DNA; that DNA is the result of life on Earth, rather than its origin. One intriguing pre-DNA possibility is what's called the RNA world hypothesis. Today, life on Earth involves enormous chemical specialization: DNA carries information, and proteins do the work that we call metabolism, primarily catalyzing chemical reactions so that our bodies readily convert sugars into usable energy and use this efficiently to perform cellular bodily functions. But there's strong evidence that the first life on Earth involved chemical multitasking by another key life molecule: ribonucleic acid, or RNA. Today RNA is often introduced in high-school biology class as DNA's less glamorous sidekick, working as a messenger to shuttle information from DNA to ribosomes, the cell's protein manufacturing centers. However, DNA and RNA are very similar. Like DNA, RNA is a string of four different kinds of nucleotide building blocks, the exception being that RNA is a single rather than a double chain, that it uses a slightly different sugar in its molecular architecture, and that it substitutes the nucleotide uracil instead of thymine.

Although today RNA gets second billing to DNA, RNA is the more dexterous molecule. RNA, unlike DNA, is capable of both carrying genetic information and of getting metabolic work done. RNA is like DNA and a protein rolled into one molecule. This makes RNA a prime candidate for an all-in-one origin-of-life molecule, capable of both reproducing itself and carrying the code to guide the copying. And there's building evidence to support this view. Many of the molecular subunits of RNA are found in carbonaceous chondrites or could have formed through chemical reactions in the Earth's primordial oceans or atmosphere. Many viruses today use only RNA as their genetic molecule, and RNA is the core of ribosomes, our cellular subunits that are factories for protein manufacturing and are known to be older than LUCA, or last universal common ancestor, the hypothesized single-cell ancestor species of all animal life on Earth. As a result, one of the dominant theories on the origins of life now being explored is that our earliest molecular ancestors were all RNA—a form of life that, Benner notes, were we to discover it today on another planet, we'd consider alien.

LIFE AS A COSMIC CONTINUUM

As a consequence of the challenge of defining life in a cosmic context, in one of the most interesting rethinkings of the Stardust Revolution, the very question of the definition of life is being turned on its head. Our difficulty in defining life isn't the problem; it's the point. Life cannot be understood as something separate from either the rest of matter or the rest of time. It can be understood only as part of a greater continuum, just as someone's individual existence can be understood only in a genealogical context. The difference here is that this new vision challenges long-held notions of an absolute divide between "life" and "nonlife." However, in the latest new way of seeing in the Stardust Revolution, what becomes evident is that there's no divide, but there is a new spectrum not of light but of life. "If the origin of life is seen as the evolutionary transition between the non-living and the living, then it is meaningless to attempt to draw a strict line between these two worlds," says Mexican evolutionary biologist Antonio Lazcano, one of the world's leading thinkers on the origins of life on Earth. "The appearance of life on Earth should, therefore, be seen as an evolutionary continuum that seamlessly joins the prebiotic synthesis and accumulation of organic molecules in a primitive environment, with the emergence of self-sustaining, replicative chemical systems capable of undergoing Darwinian evolution."

It's a view that's been slowly building against the tide of dominant thinking, from the very beginnings of the Stardust Revolution. By the end of the nineteenth century, the molecules formerly thought to be produced only by life—sugars, urea, and other carbon- and nitrogen-based compounds—were seen also to originate abiotically, without life. Near the end of his life, Charles Darwin reflected on this chemical revolution and its relationship to the nature of life itself: "Though no evidence worth anything has as yet, in my opinion, been advanced in favor of a living being, being developed from inorganic matter," Darwin wrote privately to a colleague in 1882, "yet I cannot avoid believing the possibility of this will be proved some day in accordance with the law of continuity. I remember the time, above fifty years ago, when it was said that no substance found in a living plant or animal could be produced without the aid of vital forces."

Today, Darwin's notion of "the law of continuity" has been extended from life on Earth to the stars. In looking to traverse this terrain tying atoms to life, stardust scientists see that stars, in their essence, though not "alive," are biological. This might seem like an outrageous statement. After all, for centuries stars have been the purview of astronomers and physicists. But the central realization of the Stardust Revolution is this: the stars are our direct ancestors. The notion that we are stardust is often taken as an esoteric physical one, in the sense that we're made of the elements formed in stars but somehow still separate from them. In fact, our nature is far more stellar than this. We embody the stuff of stars, and our essential nature is informed by them; we inherit information from them. Life's structure is determined by stars. They don't just forge all the elements heavier than hydrogen and helium; the ratio of these elements determines the nature of life. Stars provide not just the energy for life but also the blueprint. We are CHNOPS beings (named after the six core elements of terrestrial life: carbon, hydrogen, nitrogen, oxygen, phosphorus, and sulfur) not because of the nature of the Earth, but because of the nature of stars. As Steve Benner puts it, "The human genome is nothing more (and nothing less) than a collection of chemical structures, recording how carbon, oxygen, nitrogen, hydrogen and phosphorous atoms are bonded in natural products directly responsible for inheritance."

When it comes to elementally differentiating life from anything else in the universe, it's not a question of absolute differences but of relative abundances. One of the most striking examples of this is that your body's elemental composition is just a bunch of hydrogen gas away from being a star. Here's why: the elemental composition of the Sun and the rest of the Solar System is a reflection of the elemental composition of the solar nebula, or cosmic birth cloud out of which the Sun and its solar siblings formed some 4.567 billion years ago. To see just how closely life and the Sun are atomically related, all you have to do is balance out our levels of hydrogen and make what astronomers call devolatized solar abundances (that is, what remains after you boil away the most volatile elements, hydrogen and helium). Then, elementally we see that we're cosmic twins with very similar relative amounts of such key elements as carbon, oxygen, and calcium.

This stellar ancestry shapes the way astrobiologists think about the question "If we find other cosmic life, to what extent will we see ourselves reflected in this cosmic mirror?" One imagining of this encounter is depicted in the classic 1979 science-fiction movie *Alien*. The crew of a spaceship returning to Earth with its cargo of ore mined from an asteroid receives a distress signal; soon, however, they have reason to send their own distress signal. They inadvertently stumble across, and become hosts to, a particularly unwelcome and gruesome visitor who the crew, and moviegoers, quickly realize is very much alive. But rather than being carbon-based, this alien is silicon-based, giving it rock-hard, bullet-deflecting skin. Inspired in part by movies such as *Alien*, many astrobiologists in the past several decades have warned against limiting our view of what life is, and what we search for elsewhere in the cosmos, to what we know on Earth. In *The Limits of Organic Life in Planetary Systems*, Benner and the other authors argue that we should search in such a way that we'd be able to detect silicon-based life; life that "eats," or gets energy from, rocks; life that passes on hereditary information in chemicals other than DNA or RNA; and life that uses a solvent other than water, such as a mixture of ammonia and water—what we on Earth would consider a powerful oven cleaner.

While such alternative forms of life might well exist, the cosmic evidence so far points to some potential common cosmic family traits. If we encounter other cosmic life, we'll be meeting not aliens but distant cosmic cousins with whom we share characteristics rooted in cosmic ecology. The laws of physics and chemistry are universal, and so might be the laws, or at least the broad rules, of universal life. The Stardust Revolution has brought us to the edge of seeing whether, just as with the Copernican and Darwinian Revolutions, we are not exceptional but rather are part of the warp and weave of creation.

Many astrobiologists believe that there might be a fundamental universal nature to biochemistry. The evidence builds from atoms up to the molecules of life. The four most common life elements on Earth—hydrogen, oxygen, carbon, and nitrogen—are also the most abundant elements in the universe, not counting helium, an element that rarely bonds with others. When these atoms bond, they do so in characteristic fashion throughout the cosmos. And from this bonding, one key, remarkable pattern emerges: cosmic chemistry is built around carbon chemistry. Of the more than fifty cosmic molecules known to date with more than six atoms, all are built on a carbon framework. Prior to the dawn of astrochemistry, we thought of carbon chemistry as the chemistry of life. Now we know that carbon chemistry is the chemistry of the universe. Our carbon nature isn't some terrestrial outlier but is rather a reflection of the carbon-based molecular nature of the cosmos. Silicon is carbon's neighbor on the periodic table and has similar bonding characteristics, but, *Alien* notwithstanding, complex silicon-based molecules haven't been observed in interstellar space.

It's even possible that our protein structure and genetic code will share fundamental commonalities with other cosmic life. Stardust scientists have discovered that the twenty amino acids used by life on Earth are far from a random cosmic sample. The amino acids used by all Earth life are those that are energetically most easily synthesized and that similarly represent the greatest possible diversity of size and chemical charges. In other words, the evolution of life on Earth sampled from the smorgasbord of available amino acids, thus achieving maximum diversity while taking the biggest helpings from the largest and easiest-to-access dishes at this cosmic potluck. Finally, abundant evidence from organic materials found in meteorites and from astrochemistry experiments shows that many fundamental cosmic chemical reactions result in hollow, spherical molecular balls that resemble cells. Thus, the ingredients and chemical pathways for carbon-based cellular life have a rich cosmic natural history, one that might play out similarly elsewhere in the cosmos.

This is not to say that all cosmic life will have the identical chemical nature of terrestrial life in detail, but rather that we should search for certain fundamental chemical patterns. This is what leading astrobiologist and NASA scientist Chris McKay calls "the Lego Principle" in the search for life. Just as with the popular children's toy, cosmic life is assembled from a limited set of chemical building blocks. But what's critical in spotting life, as with any interesting Lego® structure, is the biological pattern with which the blocks are assembled. On Earth, the three main molecules of life—DNA, proteins, and polysaccharides—are all polymers, chains of simpler building blocks. Every protein—including titin, the largest protein and an awesomely complex interwoven chain of about thirty thousand amino acids that puts spring in your muscles—is built Lego-like from a bucket of just twenty different amino acids. Most impressive, all terrestrial life's incredible diversity—from the snake's pointy tongue to a starfish's wonderful suction cups—is encoded in just five nucleotide bases: adenine, guanine, cytosine, uracil, and thymine; a list biochemists reduce down to the symbols A, G, C, U, and T. Out of this five-letter chemical alphabet emerges the complex polyglot language of life. "Different life forms are likely to have different patterns," notes McKay, but "a complete analysis of the relative concentration of different types of organic molecules might reveal a pattern that is biological even if that pattern does not involve any of the familiar biomolecules." In other words, life elsewhere in the cosmos might be pure love, as the NASA engineer I spoke with during the coffee break suggested, but love has its ways, and by looking for these patterns, astrobiologists will be able to recognize the kiss of life.

In the Stardust Revolution, though, many astrobiologists believe that the debate over the first other cosmic life won't be about cellular biochemistry but rather will be about planetary chemistry. Our first glimpse of life elsewhere in the cosmos—of a possible second cosmic genesis—will come not through a microscope but through a telescope. And the debate will be about whether what we're seeing is an entire world with a heartbeat.

BUNSEN AND KIRCHHOFF'S GIFT

Today, along the hallways from Alex Dalgarno's office at the Harvard-Smithsonian Center for Astrophysics, are the offices of a new generation of bright-eyed astronomy professors and ambitious graduate students who've come of age in a different cosmos from Dalgarno's—and none more so than tall, energetic, and affable Harvard exoplanet astronomer David Charbonneau. As a teenager, Charbonneau recalls sitting in his driveway in suburban Ottawa, Canada, reading Stephen Hawking's *A Brief History of Time* and being inspired to pursue a career in theoretical astrophysics. When he arrived at Harvard to begin his PhD in 1996, it was at a moment in the Stardust Revolution when astrophysicists were seeing the history of time with new eyes. The previous year marked the discovery of the first exoplanet, and this changed everything. Charbonneau and his generation came of intellectual age in an era when exoplanets were the stuff of textbooks, courses, and careers, rather than of speculation and fruitless striving, as had been the case for a previous generation of astronomers. Their inheritance is a new cosmos, one populated with boundless other worlds. Unburdened from wondering *if* there are exoplanets, the energy and time of a new generation of astronomers is dedicated to exploring what exoplanets are like and how they form. They are a generation who senses that George Ellery Hale's vision is within their grasp: to understand the astrophysical conditions that lead to Earth-like planets. Not just Earth-sized or rocky planets, but *living* planets—the great nexus of astronomy and biology.

"What drives us is the hope that we can be part of some big change in how we think about the universe," says Charbonneau, a member of Harvard's Origins of Life Initiative, who in 2000 forged his place in exoplanet history by discovering the first transiting exoplanet. "For the first time in human history we don't just have to speculate about whether there's life elsewhere in the universe. Now we have the technology to actually look for Earth-like planets, and when we find them we'll be able to study them for signs of life."

Today in the Stardust Revolution, the search for other cosmic life has shifted from the late twentieth-century hope of discovering some extreme—by Earth standards—life eking out a living on some other body in our Solar System, to that of finding another living planet, even a verdant cosmos. The names of the proposed next generation of space-based telescopes tell the story of their intent: Darwin, Gaia, and the less prosaically named Terrestrial Planet Finder. These are telescopes whose names speak as much to biology as to astronomy. Their goal isn't to search for something "other" and different but to look for something very similar to us. We stand potentially on the edge of a great era of cosmic comparative biology. Charles Darwin's revolutionary insights into the nature of life on Earth were inspired by his circumnavigation of the world as the naturalist aboard the HMS *Beagle*. Now a new generation of telescopes could enable us to travel the inlets and islands of the Milky Way in search of other worlds with which we can compare our own. It will be through seeing ourselves reflected elsewhere, as if the night sky is a great puddle into which we can look and not just see our reflection but also compare it with other views of life, that we will better know ourselves.

The first signs of life on a distant exoplanet won't be grainy images of alien life flashing across our television screens. The stories carried by light from an exoplanet's atmosphere will be our finger on the pulse of a world light-years away. It's in these exoplanets' atmospheres, the oh-so-distant film of gas incubating the planet, that stardust scientists are exploring for the first signs of another planet with a heartbeat. A century and a half after Robert Bunsen and Gustav Kirchhoff realized that an element's light fingerprint was the same on Earth as it is in the stars, spectroscopy—the gift they gave humanity—might well carry the first news of another living planet. Starlight passing through an exoplanet's atmosphere will enable the spectroscopic identification of the gases, including "biosignature gases," a mix that's characteristic of life. Long before we directly glimpse an exoplanet's surface, we will know it by its atmosphere.

While from ground level we think of the evidence for life as blooming, hopping, or swimming in forests, fields, and seas, astronomers have long looked up to a planet's atmosphere for signs of life. Indeed, the Gaia hypothesis—the notion that life doesn't actually live on a planet but that Earth, or any other living planet, is an interactive, integrated complex system of forces that *are* life—was inspired by James Lovelock's efforts to develop remote planetary life–detection systems. In the mid-1960s, Lovelock worked at the bioscience division at NASA's Jet Propulsion Lab in Pasadena, California, where his job was to figure out how to determine whether other planets in our Solar System, particularly Mars, harbored life, without having to go there to search at ground level. Lovelock realized that a planet's atmosphere shapes, and is shaped by, life. He made the space-age leap that, in looking for alien life, rather than limiting the notion of "life" to cellular biochemistry, it was possible to identify a living planet by the composition of its atmosphere.

In a way similar to the broadening chemical vision of life's essential nature, Lovelock extended an understanding of what life does in a planetary context. One of the things it does is exchange gases. The key to spotting a living planet, said Lovelock and his colleague Dian Hitchcock, was to see a chemical disequilibrium among the mix of gases that make up a planet's atmosphere. By this they meant a mix of gases that, if left alone, would either react with one another or be broken apart by their star's light, but that in either case would disappear. Something, or someone, had to be maintaining this chemical disequilibrium. In the Earth's case, Lovelock argued, an alien civilization would be able to tell that Earth was living based on the imbalance of oxygen and methane in our atmosphere, though it wouldn't be clear what kind of life was off-gassing in this way. By volume, Earth's current atmosphere is about one-fifth oxygen, an amount that is an estimated eight times greater than what would exist if green plants were to stop pumping it out. A living planet has an atmosphere that says this must be so.

That might be the case in theory, but once the first exoplanet was discovered, few astronomers believed it would ever be possible to actually capture the tenuous light from the atmosphere of an exoplanet dozens or hundreds of light-years distant. That changed in 2002, when David Charbonneau first captured light from an exoplanet's atmosphere, peering for the first time into the sky of a world around another sun. He did this by pioneering a clever technique that's become the bread and butter of exoplanet-atmospheric scientists. When an exoplanet transits its star, passing in front of the star as viewed from Earth, the exoplanet's presence is inferred from the characteristic temporary dip in the brightness of the starlight. However, some of the star's light isn't blocked but rather shines through the exoplanet's atmosphere, picking up the exoplanet atmosphere's light fingerprint. At this point, the light tells the tale of both the chemical makeup of the star and the exoplanet's atmosphere. To extract only the exoplanet's atmospheric signature, the astronomer subtracts the star's spectrum (captured when the exoplanet is behind its star) from that of the combined star and exoplanet. Charbonneau used the Hubble Space Telescope to capture the first view of a Jupiter-sized exoplanet's atmosphere. What he recorded were the telltale doublet absorption lines in the yellow part of the spectrum, the light fingerprint of element number 11, sodium. These were the same lines that had first enthralled Robert Bunsen in his Heidelberg laboratory, and the first that Gustav Kirchhoff identified in the Sun's atmosphere. And here, once again, was sodium—the sister element of chlorine that forms the salt used at every table on Earth—shining forth from high in the sky of a faraway world, one that suddenly wasn't so alien but was joined with us in its salty essence.

In the decade since that first exoplanet atmospheric identification, we've moved from the realm of the implausible to the particular and detailed. Now the study of exoplanet atmospheres is its own realm of exploration, with specialists and a wave of upcoming younger researchers imagining how to look into alien skies. Using space-based telescopes and viewing from Earth, exoplanet sky watchers have looked into more than two dozen exoplanet atmospheres, measuring the presence of familiar molecules, including water, carbon dioxide, carbon monoxide, methane, ammonia, and helium. More than just taking an atmospheric inventory, this research is providing the first insights into the amazing realms of exoplanet weather, climate, and even geology, as well as giving us the ability to deduce an exoplanet's surface geology from its atmospheric mix. The ultimate goal of all this atmospheric searching isn't to discover distant geology but to find hints of the life on another world. As exoplanet scientists finesse their art, it's in the hope that one day soon they'll be able to capture light from a planet in its star's habitable zone, a planet with liquid water, and to probe that atmosphere for biosignatures—the telltale gases of life.

In preparing for this, we're undertaking one of the most intriguing aspects of the Stardust Revolution: thinking of Earth as an exoplanet. It is the final Copernican step, to relinquish our hold on the notion of Earth as the only living planet and to move toward seeing ourselves as being on one kind of living planet. Exoplanet scientists have gotten a good glimpse of what Earth's atmosphere would look like to observers from another solar system by examining "Earthshine." During the sliver of a crescent Moon, sunlight reflected from our home planet creates Earthshine, a glowing illumination visible to the naked eye on the larger, darker part of the Moon. In this way, the lunar surface becomes a mirror to our nature. When astronomers dissect Earthshine's infrared spectrum, in the valleys and peaks of the light curve, Earth's atmospheric mix of oxygen and ozone stands out as a beacon of what's happening below. Using a variant of this technique that looks at polarized light, astronomers have even been able to identify the presence of Earth's cloud cover, oceans, and the presence of vegetation. From Earth's perspective, oxygen and ozone as biosignatures are particularly strong indicators. We don't know of any abiotic, or nonliving, sources of oxygen in this quantity. And ozone, created when an oxygen pair is split by the Sun's ultraviolet radiation and recombines in an oxygen triplet, requires a steady source of oxygen to sustain its presence, since the same forces that create it will eventually destroy it.

However, the search for the biosignatures of another living planet isn't limited to our love affair with oxygen. Millions of species of bacteria on Earth are anaerobic—in your gut, in your compost heap, and in hot springs, they breathe gases other than oxygen. Indeed, for the majority of Earth's history, most life on Earth was anaerobic. Thus it's entirely possible that the first sister Earth we discover will be a cosmic sibling with a difference—anaerobic all the way down. In this case, such planets' distinctive atmospheric biosignatures won't be oxygen but might be a variety of anaerobically produced sulfur gases, including dimethyl sulfide, a component of cooked-cabbage and stinky-fish smell. We'll know when we've found our cosmic cousins by what they smell like.

"In the coming ten to twenty years we should have the first list of potentially habitable planets in the Sun's neighborhood," says Swiss astronomer Michel Mayor. "Making such a list is essential before future experiments can search for possible spectroscopic signatures of life in exoplanet atmospheres." However, at present, taking this spectroscopic pulse of a potentially living exoplanet is largely beyond our reach. Although astronomers have made major gains in using ground-based telescopes to capture impressive direct images of giant exoplanets, most agree that getting a high-resolution light fingerprint of an exoplanet equivalent of our small blue dot will require a new generation of specialized space-based telescopes. These telescopes will be able to block out a star's intense light in order to see the million-times dimmer reflected light of a small, rocky planet in the star's habitable zone. Yet all these proposed telescopes—the European Space Agency's Darwin and Gaia missions and NASA's Terrestrial Planet Finder—are either on hold or have been canceled. As with Hubble, Spitzer, and Kepler, a new generation of space-based telescopes designed to discover and explore living exoplanets will lead to yet another new way of seeing the cosmos both outside and within ourselves.

Presidents have talked of going to the Moon and to Mars, but there's yet to be a president or prime minister who's talked passionately about embracing our cosmic genealogical search. In the United

States, the culture wars over terrestrial evolution still stifle public acknowledgment of NASA and American scientists' leading twenty-first-century role in piecing together the story of cosmic evolution. At a seminal moment in the Stardust Revolution, we stand as if before a box containing old family letters—and we hesitate. The science of the Stardust Revolution has changed our view of the cosmos, but perhaps we stand unwilling to change, and to embrace, a new view of ourselves.

AN ANCIENT VIEW WITH STARDUST EYES

I get a different view of humanity's relationship to the stars when I find myself standing on the roof of the University of Guanajuato in central Mexico, peering out at the stars sparkling in the inky sky above this bowl-shaped city that spills up the surrounding hillsides. Guanajuato is best known for another kind of sparkling white light. It's referred to as the city that silver built because of the local silver mines—among the world's richest—which over the past four centuries have produced an astounding one-twentieth of all the world's silver. On this night, I look out over the city's twinkling, vibrant nightscape with stardust eyes and know that Guanajuato is ultimately the city that supernovas built. Every atom of every one of the millions of ounces of precious metal hauled back-breakingly from the area's labyrinthine underground mines was forged before the formation of the Earth in the seconds-long death throes of a giant star, a star whose light is long gone but whose body remains here today, glinting underground in the light from miners' helmets.

Through a small telescope, at eighty times magnification, the staff astronomer takes me on a visual tour of the heavens that traces the great arc of human discovery. We start with a bright point of light to the west, one that hangs midway up the inverted bowl of sky. As I focus the view, this point of light resolves into what is clearly a disk—not a point, but a planet: Venus. As I watch, Venus gradually moves from the center of vision toward the edge—an illusion, of course; it's not Venus that's moving so quickly, but me. I'm experiencing the Earth's rotation, my planet's rotation, as I watch another planet. Then my celestial guide aims the telescope at a point of light farther up the sky, at another planet, Jupiter. I am seeing not just planets but a solar system. Around Jupiter, like a cluster of dancers forming a single plane of orbit, are three of this giant planet's moons. This is the same view that for Galileo solidified the Copernican Revolution: if one looks with greater acuity, neither the Earth nor the Sun is the center of all, but we see that moons orbit other bodies. My guide adjusts the focus and asks me to again look closely at Jupiter. I see pencil-thin lines, the stripes of Jupiter's atmosphere. I'm looking not just up through our sky but into another, just as astronomers now plan to take the light fingerprint of a distant Earth-like planet's atmosphere.

My guide turns the telescope to the west and up toward three stars in a horizontal line that form Orion's belt, which holds up invisible clothes across the waist of the celestial hunter. Above the belt is a bright red star, Betelgeuse. I tell my teenage son and daughter, whom I've coaxed to come along, that what appears as a point of red light is in fact a near-death supergiant star, so large that, were it in the place of our Sun, its pulsing outer atmosphere would swallow us where we stood. When it eventually explodes as a supernova, it will seed the Milky Way with silver and gold atoms that might one day be dug up as cosmic treasure by some future Milky Way civilization on a planet yet to be born. My kids both shrug and turn to look down at the brighter, more inviting lights of the luminous city below. I think of Mount Wilson astronomer Paul Merrill—the man who first glimpsed that stars are the sites of the origin of the elements—and his comment that, with stars, "the contrast between the apparent and the real is the most stupendous in all human experience."

I turn again to look through the telescope, which is now positioned to look not at a dying star but at fiery, compact newborn stars. Below Orion's belt is a vertical line of three points of light that form Orion's sword; locally also known as the Three Marys, for the three

Marys who, according to the Four Gospels, accompanied Jesus's mother to his grave. It's the middle point of light that my guide wants me to see: part of the Orion Nebula. Through the eyepiece I see not just stars but also something so unlike what we usually see in the star-filled sky: a hazy cloud. In a pocket of the cloud are four clustered blue-white stars, youngsters whose nascent energy is blowing away their natal cocoon of cosmic gas and dust. I think about William Herschel and other nineteenth-century astronomers first seeing these great clouds, or nebulae, and wondering how they fit into the scheme of the eternal heavens. I think about Charles Townes, who in the late 1960s first saw these clouds for what they are: water heavy.

Finally, the telescope is turned nearly straight up, and before looking I ask my guide what I'll be seeing. He does not reply. As I put my eye to the eyepiece, I just glimpse his smile. The light is so bright that I need to squint: the Moon. The focus is along the border of light between the half-Moon's lit and shadowed sections. What I see makes me think not of astronomy but of geography. From a central Mexican rooftop, I see a distant terrain of massive, circular craters, their central ejecta dimples clearly visible, some craters with smaller ones in turn dimpling their interiors. Here in the Moon's pockmarked face is the story of the tumultuous, collisional formation of our Solar System, the great bombardment erased from Earth's surface by eons of tectonic recycling but still there for anyone to see on a clear night. A bombardment that brought the products of stars like Betelgeuse to a new-formed Earth.

I understand why, for my teenage children, looking down at Guanajuato's sparkling lights is more intriguing than squinting up at the stars. As my daughter says, they're so small and far away. Just as nineteenth-century philosopher Auguste Comte said it was impossible that we'd know a star's nature, it's also hard to imagine that stars have anything to do with us—with our daily challenges, our hopes, our loves, our sorrows, our futures. My daughter is engrossed with the present and the immediate, with the knowledge that while she's up here she's missing countless new Facebook® postings. Observational astronomy is in many ways about the past. The starlight we see in the night sky has been traveling for years, centuries, or millennia. The photons I saw from the Orion Nebula had traveled at the speed of light for about 1,344 years. Those photons left newborn stars several centuries after the time when, 250 miles south of Guanajuato, the ancient city of Teotihuacán was one of the largest in the world.

Today, Teotihuacán remains an impressive, sprawling ancient complex built around its central monuments, two massive pyramids; the larger of the two is the biggest pyramidal structure outside of Egypt. When the Aztecs took control of this great city, they named the pyramids in honor of their central spiritual symbols, creating the pyramid of the Moon and the larger pyramid of the Sun. The Aztecs felt a visceral connection with the Sun and stars. Over the centuries, they sacrificed tens of thousands of slaves and captured warriors to ensure the rising of the morning Sun. To them, the life force of human blood and that of the Sun were one and the same. These blood sacrifices had deep social and religious significance, both reflecting and reinforcing a cosmic connection. As Mexican poet Octavio Paz wrote, "To the ancient Aztecs the essential thing was to maintain the continuity of creation; sacrifice did not bring about salvation in another world, but cosmic health."

From these ancient times, we've come full circle in our relationship with the stars. For centuries in Western cultures, we thought the stars were utterly "other." Now we know they are like us—they're born and they die, and in the process, they change the little part of cosmic time and space that is theirs. They are not so different from us; we are not so different from them. The iron that gives our blood its red hue is the same iron glowing in the Sun's atmosphere—iron atoms ultimately forged in dying stars.

In the Stardust Revolution, we've come to know that our connection with the stars isn't really about the past. Our Solar System is one generation in an ongoing cycle of solar system birth and death. Stars are our ancestors, and more than five billion years from now, after our Sun has swollen in old age and consumed the Earth in its atmosphere, every atom of you and me will begin a great journey to form a new generation of solar systems. We're not just the children of stars; we're also their foreparents. Will life reemerge anew in this great process of cosmic ecology? Has it already done so, elsewhere in the cosmos? For all the monumental insights of the Stardust Revolution, this is the question that still shimmers like a giant glistening question pinned to a cosmic curtain, the audience waiting for the curtains to part, for the show to begin.

It's also the question hanging in the air at the end of an evening plenary session on the possible abundances of Earth-like planets at the Extreme Solar Systems II conference in Jackson Hole, Wyoming, in the fall of 2011. I stand in the almost empty hall talking with Bill Borucki. Overhead, the Kepler Space Telescope he pioneered was in the midst of staring intently, watching for the tiny shadows of other possible Earth-like planets to cross the face of distant stars. During the question period, I'd stood up and said that most people weren't really interested in the absolute abundance of possible alien Earths—whether 5 or 25 percent of the billions of Sun-like stars in our galaxy had Earth-like planets in a habitable zone—but rather when he thought there was even just *one* other such world in the entire Milky Way. Borucki stiffened and defensively replied that he was an astronomer, not an astrologer. In front of his professional peers, Borucki wasn't about to provide an answer for which there was absolutely no concrete evidence.

After the session, in the emptying room, Borucki tells me the story of a letter he'd received that helps me understand the emotional context of his response. The letter was from a dying woman who wrote that before she died, she wanted to know whether there was life elsewhere in the cosmos. A fellow American, she believed that Borucki had already discovered the answer but was conspiratorially withholding this deep truth. She'd asked a question that wells up in many of us when we're faced with the vision of our end or at a point in our lives when we find ourselves feeling the endless mystery of our existence. Our extreme genealogical searching isn't just about tracing our origins; it is also about how this cosmic family tree connects us with others. For it's in this connection that we find meaning and belonging. Her question was as much about her—about us—as it was about any cosmic cousins whom we might encounter.

To be at home in the cosmos isn't just to know our stellar origins but to find our cosmic kin. In this moment of the Stardust Revolution, we are as in facing death, looking at the deepest truths of who and what we are, so that we might sink into the wonder of what it all means. Something in us deeply wants to connect with a larger, living cosmos, not just for what it will tell us about what's out there, but also what it will tell us about what's in each of us, about what each of us *is*. I, like the dying woman who wrote to Bill Borucki, would like to share in this next step before I return to stardust.

ACKNOWLEDGMENTS

My favorite description of the writing process is that it's a long journey in a small room. Researching and writing this book certainly was a long journey, one that I would never have completed without the generosity and support of many others.

I'm indebted to radio astronomer Jan Hollis for planting the seed for this book with his compelling comment that "we now observe a universal prebiotic chemistry."

The book's outline took shape in the fall of 2008 during my time as the writer-in-residence at the Kavli Institute for Theoretical Physics, University of California–Santa Barbara. I'm grateful to the Kavli and its staff for providing the time and space for this book to take off. While at Kavli, I was able to interview many of the international participants in the Kavli's "Building the Milky Way" symposium. These interviews provided a framework on which I could build. I'm particularly indebted to Leo Blitz for saying: "You should talk to Charles Townes"; to Jennifer Johnson for introducing me to the question of the cosmic origin of the elements and the astronomer's periodic table; and to Andrew McWilliam for patiently introducing me to the lives of red giant stars and for mentioning Paul Merrill's discovery of technetium in these stars. Thank you, Priscilla Bender-Shore, for your wise counsel about being patient with the creative process.

Researching the story of our origin in the stars required that I tap into a vast repository of diverse knowledge. I'm enormously grateful to the numerous "stardust" researchers listed in the interview section who took the time to share their expertise and vision in person and by phone, who provided valuable background material, and who answered follow-up questions. Similarly, thanks to the

Astrobiology Science Conference (AbSciCon) community for sharing perennially insightful and inspiring research. Without these scientists, there wouldn't be a Stardust Revolution. I'd particularly like to thank Louis Allamandola, Lynn Rothschild, John Grula, and Robert Hazen.

I'm grateful to the scientists who generously agreed to read and comment on draft chapters and to those who supplied many of the images in the book. Getting my hands on technical tomes while living in small-town Canada was made possible thanks to the unflagging efforts of Monica Blackburn and the staff of the Almonte branch of the Mississippi Mills Public Library. Thanks to former Canada Museum of Nature colleague Bob Gault for dropping off a brown envelope at my front door with the enormously helpful February 2011 "Cosmochemistry" issue of *Elements* magazine.

Moving from idea to manuscript required lots of publishing and editorial midwifery. Thanks to agent Judy Heiblum of Sterling Lord Literistic, Inc., for pitching the project far and wide; to Prometheus Books former acquiring editor Linda Greenspan Regan for "getting" a story about the merging of evolution and astronomy; to freelance editor John Eerkes-Medrano; to Prometheus editor in chief Steven L. Mitchell; and to Prometheus assistant editor Julia DeGraf for giving the text its final polish. Thanks also to Prometheus Books designer Jackie Nasso Cooke for the awesome cover.

When the days in a small room felt long, I was particularly appreciative of writerly friends—Stephen Pincock, for his keen insight and wonderful Aussie enthusiasm, and Chris O'Brien, always my favorite guy with whom to talk books. I'm grateful to friends who provided a home away from home: in Guanajuato, Paul Marioni provided a wonderful and quiet place for me to work; while interviewing in the Bay Area, Elizabeth Cotton provided a much needed pied-à-terre.

Above all, thanks to my family: wife Rosemary Leach for sharing the creative journey and for often holding the proverbial fort while this book was at sea, and children Max and Francesca, the sparkling stardust of my life, who found humor and offered strength while I was on a long journey amid the stars. I hope this book guides you on your journeys.

A NOTE ON SOURCES

These sources outline my process of discovery and offer material for those interested in reading further.

INTERVIEWS

Interviews for this book were conducted over a three-year period and were turned into transcripts. Unless otherwise noted, all quotations in the book directly attributed to individuals are from the following interviews. The interviewees are organized in the order in which each one first appears in the book.

Lucy Ziurys, interviews with author, University of Arizona, Tucson, December 5–6, 2008.

Steve Padilla, interview with author on his visit to the 150-Foot Solar Tower on Mount Wilson, California, April 14, 2011.

Geoffrey Burbidge, phone interview with author, November 19, 2008.

Antonio Lazcano, interview with author, Woods Hole, Massachusetts, October 13, 2010.

Michael Werner, interview with author, Spitzer Space Science Center, Pasadena, California, April 14, 2011.

Charles Townes, interview with author, University of California–Berkeley, November 25, 2008.

Bob Freund, interview with author, Kitt Peak, Arizona, December 5, 2008.

Scott Sandford, interview with author, NASA-Ames, Moffett Field, California, November 26, 2008.

Richard Herd, interview with author, Geological Survey of Canada, Ottawa, Canada, October 6, 2011.

Michel Mayor, interview with author, Jackson Lake Lodge, Wyoming, September 13, 2011.

Gordon Walker, phone interviews with author, August 7 and September 2, 2009.

Bruce Campbell, phone interview with author, August 18, 2009.

Alan Boss, phone interview with author, August 5, 2009.

David Charbonneau, phone interview with author, August 4, 2009; interview with author, Jackson Lake Lodge, Wyoming, September 17, 2011.

Geoff Marcy, phone interview with author, August 4, 2009; interview with author, Jackson Lake Lodge, Wyoming, September 16, 2011.

Bill Borucki, interview with author, Jackson Lake Lodge, Wyoming, September 13, 2011.

Natalie Batalha, interview with author, Jackson Lake Lodge, Wyoming, September 14, 2011.

Alexander Dalgarno, interview with author, Harvard-Smithsonian Center for Astrophysics, Cambridge, Massachusetts, October 12, 2010.

Steve Benner, interview with author, AbSciCon 2010, Houston, Texas, April 28, 2010.

PART 1: BORN OF STARS

CHAPTER 1. THE STARDUST REVOLUTION

This introductory chapter draws on a variety of books and articles that are alternately resources and inspirations. Two classic, inspiring, and often poetic books of the Stardust Revolution are Carl Sagan,

Cosmos (New York: Random House, 1980), the companion to the influential, eponymous TV series; and Hubert Reeves, *Atoms of Silence: An Exploration of Cosmic Evolution*, trans. Ruth Lewis and John Lewis (Cambridge, MA: MIT Press, 1984). More recent additions are Armand Delsemme, *Our Cosmic Origins: From the Big Bang to the Emergence of Life and Intelligence* (Cambridge: Cambridge University Press, 1998); and David Grinspoon, *Lonely Planets: The Natural Philosophy of Alien Life* (New York: HarperCollins, 2003).

Two excellent general-reference astronomy resources that I always liked to have within reach are Frank Shu, *The Physical Universe: An Introduction to Astronomy* (Mill Valley, CA: University Science Books, 1982); and Ken Croswell *The Lives of Stars* (Honesdale, PA: Boyds Mill, 2009)—a gem that supports my adage that the best kids' books are interesting and useful for readers age five to ninety-five.

The rise of astrobiology as a science in the United States is thoroughly recounted and meticulously documented in Steven J. Dick and James E. Strick, *The Living Universe: NASA and the Development of Astrobiology* (New Brunswick, NJ: Rutgers University Press, 2005). In the United States, the committees and publications of the National Research Council's Space Sciences Board have been critical in documenting and providing a vision for the Stardust Revolution. Two key reports, both available from the National Academies Press website, are National Research Council, *The Astrophysical Context of Life* (Washington, DC: National Academies Press, 2005); and National Research Council, *The Limits of Organic Life in Planetary Systems* (Washington, DC: National Academies Press, 2007). Two articles that have marked the emergence of the development of astrobiology as a discipline are L. J. Mix et al., "The Astrobiology Primer: An Outline of General Knowledge," *Astrobiology* 6, no. 5 (2006): 735–813; and David J. Des Marais et al., "The NASA Astrobiology Roadmap," *Astrobiology* 8, no. 4 (2008): 715–30. Putting astrobiology in historical context is William Brazelton and Woodruff Sullivan III, "Understanding the Nineteenth Century Origins of Disciplines: Lessons for Astrobiology Today?" *International Journal of Astrobiology* 8, no. 4 (2009): 257–66; and Isaac Asimov, "A Science in Search of a Subject Matter," *New York Times Magazine*, May 23, 1965, pp. 52–58. For a big-picture perspective on scientific revolutions, see Thomas S. Kuhn, *The Structure of Scientific Revolutions*, 2nd ed. (Chicago: University of Chicago Press, 1970).

Sources of Direct Quotations by Page

(page 21) "*We live in a changing universe . . .*" Timothy Ferris, *The Whole Shebang: A State-of-the-Universe(s) Report* (New York: Simon & Schuster, 1997), p. 11.

(page 23) "Origins *is one of the boldest challenges . . .*" Daniel Goldin, "Speech to Congress," made in testimony to US Congress on May 22, 1996, Washington, DC, NASA, HQ Historical Reference Section, https://mira.hq.nasa.gov/history/ws/hdmshrc/all/main/Blob/19263.pdf;jsessionid=F4F38152BBB70617BF4596FD849878E6?rpp=100&m=386&w=NATIVE%28%27SERIES+ph+any+%27%27Goldin%27%27%27%29&order=native%28%27DOC_DATE%2FDescend%27%29 (accessed December 6, 2011).

(page 24) "*there are compelling reasons . . .*" National Research Council, *Astrophysical Context of Life*, p. 21.

(page 27) "*should be connected to topics . . .*" Ibid., p. 21.

(page 27) "*We must move beyond the circumstances . . .*" Des Marais et al., "NASA Astrobiology Roadmap," p. 720.

(page 29) "*A complete, consistent, unified theory . . .*" Stephen Hawking, *A Brief History of Time* (New York: Bantam Books, 1988), p. 169.

(page 31) "*The surface of the Earth . . .*" Sagan, *Cosmos*, p. 5.

(pages 33–34) "*I grew up with the erroneous notion . . .*" Fred Hoyle, *Home Is Where the Wind Blows: Chapters from a Cosmologist's Life* (Mill Valley, CA: University Science Books, 1994), p. 289.

CHAPTER 2. A STAR'S FINGERPRINT

Two key reference books on the history of stellar spectroscopy, or fingerprinting, are J. B. Hearnshaw, *The Analysis of Starlight: One Hundred and Fifty Years of Astronomical Spectroscopy* (Cambridge: Cambridge University Press, 1986); and Marcus Chown, *The Magic Furnace: The Search for the Origin of Atoms* (London: Jonathan Cape, 1999).

Sources by Sections

Looking at the Sun

The history of Mount Wilson's 150-Foot Solar Tower is recounted in Pam Gilman, "The 150-Foot Solar Tower History," http://obs.astro.ucla.edu/150_hist.html; and Larry Webster, "Daily Sunspot Drawings at the 150-Foot Solar Tower," 2007, http://obs.astro.ucla.edu/150_draw.html (both accessed August 8, 2011). You can see Steve Padilla's view from the top of the 150-Foot Solar Tower on any day here: http://obs.astro.ucla .edu/towercam.htm.

The Great Seer

The authoritative biography on George Ellery Hale is Helen Wright, *Explorer of the Universe: A Biography of George Ellery Hale* (New York: E. P. Dutton, 1966). An excellent visual summary of Hale as a telescope-building titan is the documentary film *Journey to Palomar, America's First Journey into Space*, directed by Robin Mason and Todd Mason (2008). Hale's own thoughts on the intersection of astronomy and evolution can be found in his book *The Study of Stellar Evolution; An Account of Some Recent Methods of Astrophysical Research* (Chicago: University of Chicago Press, 1908).

Out of Mystery

Auguste Comte's thoughts are taken from his book *The Positive Philosophy of Auguste Comte*, translated and condensed by Harriet Martineau (New York: Calvin Blanchard, 1858). An overview of Comte's life and thoughts are at the Stanford Encyclopedia of Philosophy, s.v. "August Comte," http://plato.stanford.edu/entries/comte/ (accessed December 12, 2010).

Bunsen's Burnings and Mystery of the Fraunhofer Lines

Robert Bunsen and Gustav Kirchhoff's discovery of spectroscopy and its application to starlight draws from the following sources: Henry Crew, "Robert Wilhelm Bunsen," *Astrophysical Journal* 10, no. 5 (1899): 301–305; George Lockemann, "The Centenary of the Bunsen Burner," *Journal of Chemical Education* 33 (January 1956): 20–22; Mary Elvira Weeks, "The Discovery of the Elements XIII: Some Spectroscopic Discoveries," *Journal of Chemical Education* 9 (August 1932): 1413–34; Dietmar Seyferth, "Cadet's Fuming Arsenical Liquid and the Cacodyl Compounds of Bunsen," *Organometallics* 20 (2001): 1488–98; Gustav Kirchhoff and Robert Bunsen, "Chemical Analysis by Observation of Spectra," *Annalen der Physik und der Chemie* 110 (1860): 161–89; Gustav Kirchhoff, *Researches on the Solar Spectrum and the Spectra of the Chemical Elements*, trans. Henry Roscoe (Cambridge, London: Macmillan, 1862); and Sam Jayakumar, *Splendor of the Spectrum*, http://www1.umn.edu/ships/modules/chem/spectroscope.pdf (accessed December 10, 2010).

Order in the Heavens

Hearnshaw's and Chown's books, listed previously, provide a good overview of the impact of the discovery of spectroscopy on nineteenth-century and early twentieth-century astronomy. In addition, see Agnes Clerke, *A Popular History of Astronomy during the Nineteenth*

Century (Edinburgh: A. & C. Black, 1885); and several articles on Sir William Huggins, the first and most enthusiastic stellar spectroscopist: W. W. Campbell, "Sir William Huggins, K.C.B., O.M.," *Publications of the Astronomical Society of the Pacific* 22, no. 133 (October 1910): 148–63; Barbara Becker, "Celestial Spectroscopy: Making Reality Fit the Myth," *Science* 301 (September 5, 2003): 1332–33; and Barbara Becker, "Eclecticism, Opportunism, and the Evolution of a New Research Agenda: William and Margaret Huggins and the Origins of Astrophysics" (PhD diss., Johns Hopkins University, 1993).

A Stranger in the Stars

An excellent overview of articles about and by Paul Merrill is listed on the Bruce Medalists website at http://www.phys-astro.sonoma.edu/brucemedalists/merrill/index.html (accessed March 13, 2012). My overview of Merrill was compiled from the following, along with a visit to the Reyn Restaurant where he enjoyed his daily lunch: A. H. Joy, "Paul Willard Merrill, 1887–1961," *Publications of the Astronomical Society of the Pacific* 74, no. 3 (1962): 41–43; George W. Preston, "Mount Wilson Observatory: Contributions to the Study of Cosmic Abundances of the Chemical Elements," in *Origin and Evolution of the Elements: Carnegie Observatories Astrophysics Series*, vol. 4, ed. Andrew McWilliam and Michael Rauch (Cambridge: Cambridge University Press, 2004) pp. 1–7; Alan Sandage, "The Mount Wilson Observatory: Breaking the Code of Cosmic Evolution," in *Centennial History of the Carnegie Institution of Washington*, vol. 1 (Cambridge: Cambridge University Press, 2005); Hale, *Study of Stellar Evolution*; Olin C. Wilson, "Paul Willard Merrill 1887–1961," biographical memoir, *National Academies of Sciences* (1964): 235–66; Paul W. Merrill, "From Atoms to Galaxies," *Leaflets of the Astronomical Society of the Pacific* 7, leaflet no. 349 (June 1958): 393–400; Paul W. Merrill, "Spectroscopic Observations of Stars of Class S," *Astrophysical Journal* 116 (1952): 21; Albert Stwertka, *A Guide to the Elements*, 2nd ed. (New York: Oxford University Press, 2002);

Harry J. Emeléus and Alan G. Sharpe, eds., *Advances in Inorganic Chemistry and Radiochemistry*, vol. 11 (New York: Academic Press, 1968); Charlotte Moore, "Technetium in the Sun," *Science* 114 (July 20, 1951): 59–61; Paul W. Merrill, "Technetium in the Stars," *Science* 115 (May 2, 1952): 484; and, finally, the posthumously published Paul W. Merrill, *Space Chemistry* (Ann Arbor: University of Michigan Press, 1963).

Sources of Direct Quotations by Page

(page 37) "*The tiny, twinkling stars of the night sky . . .*" Paul W. Merrill, "Stars as They Look and as They Are," *Publications of the Astronomical Society of the Pacific* 38, no. 221 (1926): 14.

(page 42) "*It is not too much . . .*" Hale, *Study of Stellar Evolution*, p. 1.

(page 42) "*We are now in a position to regard the study of evolution . . .*" Ibid., p. 3.

(page 45) "*Men will never encompass . . .*" Comte, *Positive Philosophy of Auguste Comte*, p. 132.

(page 46) "*Experiment is, of course, impossible.*" Ibid.

(page 46) "*could take place only . . .*" Ibid., p. 134.

(page 46) "*measuring angles . . .*" Ibid.

(page 46) "*if the knowledge of . . .*" Ibid., p. 133.

(page 47) "*the work for which Bunsen . . .*" Crew, "Robert Wilhelm Bunsen," p. 303.

(page 51) "*If there should be substances . . .*" Kirchhoff and Bunsen, "Chemical Analysis by Observation of Spectra," p. 189.

(page 55) "*Spectrum analysis . . . opens to chemical . . .*" Kirchhoff, *Researches on the Solar Spectrum*, pp. 20–21.

(page 55) "*If we were to go to the Sun . . .*" Becker, "Celestial Spectroscopy," p. 1332.

(page 56) "*The news reached me . . .*" Campbell, "Sir William Huggins," p. 150.

(page 60) "*a most seriously, serious scientist.*" Sandage, "Mount Wilson Observatory," p. 284.

(page 60) "*cloud over like a summer thunderstorm . . .*" Ibid., p. 285.

(page 61) "*a truly monumental mass . . .*" Wilson, "Paul Willard Merrill 1887–1961," p. 242.

(page 61) "*passionate enthusiasm for the study . . .*" Ibid., p. 240.

CHAPTER 3. THE ORIGIN OF THE ELEMENTS

Two key books that contributed to many parts of this chapter are Eric R. Scerri, *The Periodic Table: Its Story and Its Significance* (Oxford: Oxford University Press, 2007); and Marcus Chown, *The Magic Furnace: The Search for the Origin of Atoms* (London: Jonathan Cape, 1999).

Sources by Sections

Of Stars and Atoms

For a thoughtful, detailed history of the Kellogg Radiation Laboratory, with a focus on William Fowler's research, see Olivia Weaver Walling, "Research at the Kellogg Radiation Laboratory, 1920s–1960s: A Small Narrative of Physics in the Twentieth Century" (PhD diss., University of Minnesota, 2005). A much shorter historical reflection by one of the key players is Thomas Lauritsen, "Kellogg Laboratory: The Early Years," *Engineering and Science* (June 1969).

The Alchemist's Dream

A fantastic account of Newton's life as not the first great modern scientist but the last of the great magicians is Thomas Levenson, *Newton and the Counterfeiter: The Unknown Detective Career of the World's Greatest Scientist* (New York: Houghton Mifflin Harcourt, 2009).

Specific details on William Crookes were drawn from Chown, *Magic Furnace*; and Scerri, *Periodic Table*; see also "William

Crookes: A Victorian 'Man of Science,'" http://www.chem.ox.ac.uk/icl/heyes/lanthact/biogs/crookes.html (accessed December 1, 2010); and William Crookes, "Genesis of the Elements," *Chemical News* 55 (February 25, 1887).

A Recipe for Sunshine

A. S. Eddington, *Stars and Atoms* (New Haven, CT: Yale University Press, 1927); and A. S. Eddington, "The Internal Constitution of Stars," *Nature* 106 (September 2, 1920): 34–49. For background on Hans Bethe's work, see Martin Harwit, "The Growth of Astrophysical Understanding," *Physics Today* 56, no. 11 (November 2003): 38–43; and Hans Bethe, "Energy Production in Stars," *Physical Review* 55 (March 1, 1939): 434–56.

Big-Bang Atoms and Let There Be Hoyle

Three scholarly books on the discovery of stellar nucleosynthesis provided guideposts for this chapter: Simon Mitton, *Conflict in the Cosmos: Fred Hoyle's Life in Science* (Washington, DC: Joseph Henry Press, 2005); Jane Gregory, *Fred Hoyle's Universe* (Oxford: Oxford University Press, 2005); and Ken Croswell, *Alchemy of the Heavens: Searching for Meaning in the Milky Way* (New York: Anchor Books, 1995).

In addition, the protagonist tells his own engaging version of the history in Fred Hoyle, *Home Is Where the Wind Blows: Chapters from a Cosmologist's Life* (Mill Valley, CA: University Science Books, 1994).

For a historical overview of the science of the origins of the universe, see Simon Singh, *Big Bang: The Origin of the Universe* (New York: HarperCollins, 2004); and "Big Bang or Steady State? Creation of the Elements," American Institute of Physics, http://www.aip.org/history/cosmology/ideas/bigbang.htm (accessed December 9, 2011).

The original scientific articles putting forth the big-bang origin of the elements are S. Chandrasekhar and Louis R. Henrich, "An Attempt to Interpret the Relative Abundances of the Elements and

Their Isotopes," *Astrophysical Journal* 95 (1942); George Gamow, "The Origin of the Elements and the Separation of the Galaxies," *Physical Review* 74 (1948): 505–506; George Gamow, "Expanding Universe and the Origin of the Elements," *Physical Review* 70 (1946): 572–73; and A. Alpher, H. Bethe, G. Gamow, "The Origin of the Chemical Elements," *Physical Review* 73 (1948): 803–804.

Fred Hoyle's two key papers arguing for ongoing stellar nucleosynthesis are "The Synthesis of the Elements from Hydrogen," *Monthly Notices of the Royal Astronomical Society* 106 (1946): 343–83; and Fred Hoyle, "On Nuclear Reactions Occurring in Very Hot Stars. I. The Synthesis of Elements from Carbon to Nickel," *Astrophysical Journal Supplement* 1 (1954): 121–46.

Detailed historical overviews of the search for the origin of the elements are Virginia Trimble, "The Origins and Abundances of the Elements before 1957: From Prout's Hypothesis to Pasadena," *European Physical Journal H* 35 (2010): 89–109; Arno Penzias, "The Origin of Elements," Nobel lecture, December 8, 1978, in *Nobel Lectures, Physics 1971–1980*, ed. Stig Lundqvist (Singapore: World Scientific, 1992), http://nobelprize.org/nobel_prizes/physics/laureates/1978/penzias-lecture.html (accessed August 9, 2011); David Arnett, "Hoyle's Synthesis of Heavy Elements," *Astrophysical Journal, Centennial Issue* 525 (1999): 597–98; Somak Raychaudhury, "And Gamow Said, Let There Be a Hot Universe," *Resonance: Journal of Science Education* 9, no. 7 (2004): 32–43; and Donald Clayton, "Hoyle's Equation," *Science* 318 (December 21, 2007): 1876–77.

For an excellent, accessible overview of cosmic elemental abundances, see Katharina Lodders, "The Chemical in the Cosmos," *Glimpse Journal* 2, no. 4 (2009–2010): 28–37.

The Astronomer's Periodic Table

Along with the above titles, the story of the how the Burbidges, Fowler, and Hoyle arrived at a detailed explanation of stellar nucleosynthesis is drawn from two reflections by the principals:

Margaret Burbidge, "Synthesis of the Elements in the Stars: B²FH and Beyond," in *Origin and Evolution of the Elements: Carnegie Observatories Astrophysics Series*, vol. 4, ed. Andrew McWilliam and Michael Rauch (Cambridge: Cambridge University Press, 2004), pp. 8–10; and Margaret Burbidge, Geoffrey Burbidge, and Fred Hoyle, preface to George Wallerstein et al., eds., "Synthesis of the Elements in Stars: Forty Years of Progress," *Reviews of Modern Physics* 69, no. 4 (October 1997): 997–98. The original B²FH paper is Margaret E. Burbidge, G. R. Burbidge, William Fowler, and F. Hoyle, "Synthesis of the Elements in Stars," *Reviews of Modern Physics* 29, no. 4 (October 1957): 547–641. The critical information that guided their understanding of the cosmic abundance of elements is found in Hans E. Suess and Harold Urey, "Abundances of the Elements," *Reviews of Modern Physics* 28, no. 1 (January 1956): 53.

For further biographical information, see Dennis Overbye, "Geoffrey Burbidge, Who Traced Life to Stardust, Is Dead at 84," *New York Times*, February 7, 2010; "Fred Hoyle: An Online Exhibition," St. John's College, University of Cambridge, http://www.joh.cam .ac.uk/library/special_collections/hoyle/exhibition/radio (accessed August 9, 2011); and David DeVorkin, "Oral History Transcript—Margaret Burbidge," *American Institute of Physics* (1978).

Nobel Conclusions

Fowler's Nobel acceptance speech is in William Fowler, "Experimental and Theoretical Nuclear Astrophysics; the Quest for the Origin of the Elements," in *Nobel Lectures, Physics 1981–1990*, ed. Tore Frängsmyr and Gösta Ekspång (Singapore: World Scientific, 1993), pp. 172–79, "William A. Fowler—Nobel Lecture," Nobelprize.org, http://nobelprize.org/nobel_prizes/physics/laureates/1983/fowler -lecture.html (accessed August 10, 2011).

An overview of the runner-up in the discovery of the origin of the elements can be found in John Cowan and James Truran, "In Memory of Al Cameron," *Proceedings of Science* (2006): 1–8.

Sources of Direct Quotations by Page

(page 67) "*It is a rather interesting* . . ." Walter Adams, "The Interior of a Star and How It Maintains Its Life," *Scientific Monthly*, April 1928, p. 363.

(page 68) "*his theoretical and experimental studies* . . ." "The Nobel Prize in Physics 1983: Subramanyan Chandrasekhar, William A. Fowler," Nobelprize.org, http://nobelprize.org/nobel_prizes/physics/laureates/1983/ (accessed December 6, 2011).

(page 70) "*Dear Fred, After the initial* . . ." William Fowler, letter to Fred Hoyle, November 3, 1983, Caltech Archives, Fowler correspondence, box 12, folder 17.

(page 72) "*Just as the world* . . ." Levenson, *Newton and the Counterfeiter*, p. 85.

(page 73) "*In the very words selected* . . ." Crookes, "Genesis of the Elements," p. 83.

(page 74) "*If the views we* . . ." Scerri, *Periodic Table*, p. 39.

(page 75) "*In these our times* . . ." Crookes, "Genesis of the Elements," p. 83.

(page 75) "*I have in this glass* . . ." Ibid., p. 85.

(page 76) "*a genetic relation* . . ." Ibid., p. 83.

(pages 77–78) "*We may say, with certainty* . . ." Chown, *Magic Furnace*, p. 50.

(page 79) "*The problem . . . was that Eddington* . . ." Hoyle, *Home Is Where the Wind Blows*, p. 146.

(page 79) "*If the [gravitational] contraction theory* . . ." Eddington, "Internal Constitution of Stars," p. 18.

(page 80) "*If we decide to inter* . . ." Ibid.

(page 81) "*that all the elements are constituted* . . ." Ibid.

(page 95) "*It was the most rapid* . . ." Croswell, *Alchemy of the Heavens*, p. 166.

(page 100) "*the abundances of the elements* . . ." Suess and Urey, "Abundances of the Elements," p. 53.

(page 100) "*Dear Hans* . . ." William Fowler, letter to Hans

Bethe, July 10, 1956, "Hans Bethe 1949–1991," Caltech Archives, Fowler correspondence, box 3, folder 33.

(page 103) "*The grand concept of . . .*" Fowler, "Experimental and Theoretical Nuclear Astrophysics," pp. 172–79.

(page 103) "*And so God said . . .*" Paul Davies, *Cosmic Jackpot: Why Our Universe Is Just Right for Life* (New York: Houghton Mifflin, 2007), p. 139.

(page 103) "*Dear Willy . . .*" Fred Hoyle, letter to William Fowler, November 8, 1983, Caltech Archives, Fowler correspondence, box 12, folder 17.

(page 104) "*The fiction is that . . .*" Hoyle, *Home Is Where the Wind Blows*, p. 272.

(page 104) "*Scientific discovery is not really . . .*" Ibid., p. 272.

(page 105) "*When singer Joni Mitchell said . . .*" S. M. Faber et al., "Tribute to Geoffrey Burbidge (1925–2010)," *Annual Review of Astronomy and Astrophysics* 48 (August 2010).

PART 2: THE INVISIBLE UNIVERSE

CHAPTER 4. THE ATOMS OF LIFE

Sources by Sections

Darwin's Gap

I attended the October 12, 2010, meeting of the Committee on the Origins and Evolution of Life in Woods Hole, Massachusetts. The meeting agenda is available at National Academies of Sciences, Space Studies Board, http://sites.nationalacademies.org/SSB/ssb_052326 (accessed April 27, 2012). NASA astrobiologist Mary Voytek presented the "NASA Astrobiology Program Update," via video conference. Darwin's thoughts on the origin-of-life debate are recounted in J. Peretó, J. Bada, and A. Lazcano, "Charles Darwin and the Origin of Life," *Origins of Life and Evolution of the Biosphere* 39 (2009): 395–406. An inspiring article is George Wald, "The Origins of Life," *Proceedings of the National Academy of Sciences* 52 (August 1964): 595–611.

On the Origin of Life

Initially oriented by a very helpful discussion with Antonio Lazcano, this section is based on the following: Antonio Lazcano, "Historical Development of Origins Research," *Cold Spring Harbor Perspectives in Biology* 2 (November 2010): 1–16; and Iris Fry, "The Origins of Research into the Origins of Life," *Endeavour* 30, no. 1 (March 2006): 24–28.

For an overview of scientific context in which Oparin's ideas emerged, see Loren R. Graham, *Science in Russia and the Soviet Union: A Short History* (Cambridge: Cambridge University Press, 1993).

The Spontaneous-Generation Debate

The copy of Oparin's masterwork to which I referred was A. I. Oparin, *Origin of Life*, 2nd ed. (New York: Dover, 1965).

The American response to Oparin's book is taken from William Marias Malissoff, "How Did Life Begin on This Strange Planet?" *New York Times*, May 29, 1938.

For an overview of the debate, see John Farley, *The Spontaneous Generation Controversy from Descartes to Oparin* (Baltimore: Johns Hopkins University Press, 1977); and from a more technical, rearview-mirror perspective, see Horst Rauchfuss, *Chemical Evolution and the Origin of Life*, trans. Terence N. Mitchell (Berlin: Springer, 2008).

The discussion of Pasteur's role is based largely on Oparin, *Origin of Life*, pp. 19–30; John Wilkins, *Spontaneous Generation and the Origin of Life*, TalkOrigins Archive, http://www.talkorigins.org/faqs/abioprob/spontaneous-generation.html, 2004 (accessed June

2011); David Cohn, "The Life and Times of Louis Pasteur," http://pyramid.spd.louisville.edu/~eri/fos/interest1.html#Spontaneous%20Generation (accessed December 20, 2011); and Gerald L. Geison, *The Private Science of Louis Pasteur* (Princeton, NJ: Princeton University Press, 1995).

An Elemental View of Life

Stanley Miller, William L. Schopf, and Antonio Lazcano, "Oparin's 'Origin of Life': Sixty Years Later," *Journal of Molecular Evolution* 44 (1997): 351–52.

Oparin, *Origin of Life*, p. 30.

A. I. Oparin, "Introductory Address," in *Proceedings of the First International Symposium on the Origin of Life on the Earth*, ed. F. Clark and R. L. M. Synge (New York: Pergamon, 1959), p. 2.

Aditya Chopra and Charles H. Lineweaver, "The Major Elemental Abundance Differences between Life, the Oceans and the Sun," *Australian Space Science Conference Series: 8th Conference Proceedings*, National Space Society of Australia, 2009, p. 2.

T. H. Huxley, "On the Physical Basis of Life," *Fortnightly Review* (New Haven, CT: College Courant, 1869), p. 16.

Molecular Evolution

This section draws directly from Oparin, *Origin of Life*, with additional perspective gained from Miller, Schopf, and Lazcano, "Oparin's 'Origin of Life,'" pp. 351–53.

Earth in Glass

This section is reassembled based on the following: Antonio Lazcano and Jeffrey Bada, "Stanley L. Miller (1930–2007): Reflections and Remembrances," *Origins of Life and Evolution of Biospheres* 38 (2008): 373–81; James E. Strick, "Creating a Cosmic Discipline: The

Crystallization and Consolidation of Exobiology, 1957–1973," *Journal of the History of Biology* 37 (2004); and Clark and Synge, eds., *Proceedings of the First International Symposium on the Origin of Life on the Earth.*

Although briefly mentioned in this section, the transformation of physicists into biologists is a fascinating story. For an overview, see M. F. Perutz, "Physics and the Riddle of Life," *Nature* 326 (1987): 555–58; and one of the original articles exploring the topic, by the father of quantum physics, Niels Bohr, "Light and Life," *Nature* (March 25, 1933): 421–59. One of the most famous of these reflections is found in Erwin Schrödinger's lectures turned book, *What Is Life? The Physical Aspects of the Living Cell* (Cambridge: Cambridge University Press, 1943).

Harold Urey's thoughts on how Earth's astrophysical context shaped its early atmosphere are outlined in his "On the Early Chemical History of the Earth and the Origin of Life," *Proceedings of the National Academy of Sciences* 38 (1952): 351–63.

Stanley Miller's landmark paper is his "A Production of Amino Acids under Possible Primitive Earth Conditions," *Science* 117 (May 15, 1953): 528–29.

For reflection on the Miller-Urey experiment and its contemporary and historical assessment, see Jeffrey Bada and Antonio Lazcano, "Prebiotic Soup—Revisiting the Miller Experiment," *Science* 300 (May 2, 2003): 746; "Life and a Glass Earth," *New York Times*, May 17, 1953; Adam P. Johnson et al., "The Miller Volcanic Spark Discharge Experiment," *Science* 322 (October 17, 2008): 404; Jennifer Evans, "Miller-Urey Amino Acids, circa 1953," *Scientist* 23, no. 1 (2009): 72; and Eric T. Parker et al., "Primordial Synthesis of Amines and Amino Acids in a 1958 Miller H2S-Rich Spark Discharge Experiment," *Proceedings of the National Academy of Sciences*, March 21, 2011, http://www.pnas.org/content/early/2011/03/14/1019191108.full.pdf+html.

For current thinking on the composition of the Earth's early atmosphere, see Dustin Trail, E. Bruce Watson, and Nicholas D. Tailby, "The Oxidation State of Hadean Magmas and Implications for Early Earth's Atmosphere," *Nature* 480 (December 1, 2011): 79–82.

Liftoff for Exobiology

The overview perspective for this section was drawn from Strick, "Creating a Cosmic Discipline," pp. 131–80; and Michel Morange, "Fifty Years Ago: The Beginnings of Exobiology," *Journal of Biosciences* 32, no. 6 (September 2007): 1083–87.

Information on Joshua Lederberg and J. B. S. Haldane is in part from "The Joshua Lederberg Papers," Profiles in Science, National Library of Medicine, http://profiles.nlm.nih.gov/BB/ (accessed March 17, 2011); and Fry, "Origins of Research into the Origins of Life," pp. 24–28.

Lederberg's two seminal papers outlining the new science of exobiology are Joshua Lederberg, "Exobiology: Approaches to Life beyond the Earth," *Science* 132, no. 3424 (1960): 393–99; and Joshua Lederberg and Dean B. Cowie, "Moondust," *Science* 127, no. 3313 (June 27, 1958): 1473–75.

Sources of Direct Quotations by Page

(page 109) "*It is a slightly arresting notion* . . ." Bill Bryson, *A Short History of Nearly Everything* (New York: Broadway Books, 2003), p. 1.

(page 111) "*The chief defect* . . ." Peretó, Bada, and Lazcano, "Charles Darwin and the Origin of Life," p. 396.

(page 111) "*It is mere rubbish* . . ." Ibid., p. 399.

(page 115) "*landmark for discussion* . . ." Malissoff, "How Did Life Begin on This Strange Planet?" p. 64.

(page 115) "*Even in our own time* . . ." Oparin, *Origin of Life*, p. 2.

(page 116) "*Art cannot combine* . . ." Lazcano and Bada, "Stanley L. Miller (1930–2007)," p. 374.

(page 118) "*Never will the doctrine* . . ." Cohn, "Life and Times of Louis Pasteur."

(page 119) "*pit of vitalist conceptions*," Oparin, *Origin of Life*, p. 30.

(page 119) "*The problem of the origin . . .*" Oparin, "Introductory Address," p. 2.

(page 120) "*we do not hesitate . . .*" Huxley, "On the Physical Basis of Life," p. 16.

(page 120) "*It is remarkable that . . .*" William Huggins and W. A. Miller, "On the Spectra of Some of the Fixed Stars," *Philosphical Transactions of the Royal Society of London* 154 (1864): 434.

(page 121) "*The carbon atom in . . .*" Oparin, *Origin of Life*, p. 136.

(page 122) "*It must be understood . . .*" Ibid., p. 246.

(page 122) "*All these difficulties . . .*" Ibid., p. 60.

(page 122) "*However strange this . . .*" Ibid., p. 63.

(page 126) "*that experimentation on the . . .*" Urey, "On the Early Chemical History," p. 362.

(page 128) "*made chemical history . . .*" "Life and a Glass Earth."

(page 129) "*if God didn't do . . .*" Lazcano and Bada, "Stanley L. Miller (1930–2007)," p. 376.

(page 132) "*Since the sending of rockets . . .*" Lederberg and Cowie, "Moondust."

(pages 132–33) "*Twenty-five centuries of . . .*" Lederberg, "Exobiology," p. 394.

(page 133) "*The dynamics of celestial bodies . . .*" Ibid.

CHAPTER 5. DUST TO DIAMONDS

Several books and articles served as valuable background resources to provide a historical and broad view: James B. Kaler, *Cosmic Clouds: Birth, Death, and Recycling in the Galaxy* (New York: Scientific American Library, 1997); Peter Martin, *Cosmic Dust: It's Impact on Astronomy* (Oxford: Oxford University Press, 1978); Bruce T. Draine, "Interstellar Dust Grains," *Annual Review of Astronomy and Astrophysics* 41 (2003): 241–89; and, finally, the excellent, detailed, and broad perspective offered by the contributors to Daniel

Apai and Dante S. Lauretta, eds., *Protoplanetary Dust: Astrophysical and Cosmochemical Perspectives* (Cambridge: Cambridge University Press, 2010); see, in particular, the chapter by Hans-Peter Gail and Peter Hoppe, "The Origins of Protoplanetary Dust and the Formation of Accretion Disks."

Sources by Sections

The Original Dark Matter

My visit to Grasslands National Park was in July 2010. For information on this park, Canada's darkest dark-sky preserve, see "Grasslands National Park," Parks Canada, http://www.pc.gc.ca/pn-np/sk/grasslands/index.aspx.

A New Land between the Stars

Two of the key scientific papers that brought awareness to the cosmos' dusty nature are R. F. Sanford, "On Some Relations of the Spiral Nebulae to the Milky Way," *Lick Observatory Bulletin*, no. 297 (1917): 80–91; and Harold Julius Trumpler, "Preliminary Results on the Distances, Dimensions and Space Distribution of Open Star Clusters," *Lick Observatory Bulletin*, no. 420 (1930): 154–88. I learned about the "Hubble of cosmic dust" in Harold F. Weaver, "Harold Julius Trumpler: 1886–1956," *Biographical Memoirs*, vol. 78 (Washington, DC: National Academy Press, 2000), pp. 3–8.

Fred Hoyle's perspective on dust is drawn from his science-fiction book: Fred Hoyle, *The Black Cloud* (New York: Signet Books, 1957); from his biographer: Jane Gregory, *Fred Hoyle's Universe* (Oxford: Oxford University Press, 2005); and from his classic article describing stardust production: Fred Hoyle and N. C. Wickramasinghe, "On Graphite Particles as Interstellar Grains," *Monthly Notices of the Royal Astronomical Society* 124 (1962). See also N. C. Wickramasinghe, "Interstellar Grains: 50 Years On," *Journal of Cosmology* (December

2011), available online at http://journalofcosmology.com/JoC16 pdfs/INTERSTELLAR%20GRAINS%20Final.pdf (accessed May 3, 2012).

Seeing with Stardust Eyes

For an understanding of the dawn of infrared astronomy, see Monica Salomone and Lars Lindberg Christensen, *The Infrared Revolution: Unveiling the Hidden Universe* (Paris: European Space Agency, Public Relations Division, 2000); Frank J. Low, G. H. Rieke, and R. D. Gehrz, "The Beginning of Modern Infrared Astronomy," *Annual Review of Astronomy and Astrophysics* 45 (2007): 43–75; and "Discovering Infrared—The Herschel Experiment," Infrared Processing and Analysis Center, http://www.ipac.caltech.edu/outreach/Edu/Herschel/ herschel.html (accessed March 22, 2012).

For background on the Spitzer Space Telescope, see two articles by its chief scientist, Michael Werner: Michael Werner, "Spitzer's Cold Look at Space," *American Scientist* 97 (November–December 2009): 58–68; and Michael Werner et al., "First Fruits of the Spitzer Space Telescope," *Annual Review in Astronomy and Astrophysics* 44 (May 2006): 269–321.

For overview information on Spitzer, see the Jet Propulsion Laboratory's excellent website at http://www.spitzer.caltech.edu/.

The Cold and Dirty Cosmos

The relationship between dust and star formation draws on Richard B. Larson, "Historical Perspective on Computational Star Formation," in J. Alves et al., eds., *Computational Star Formation, Proceedings of the International Astronomical Union Symposium*, no. 270 (2011); Bart Bok and Edith F. Reilly, "Small Dark Nebulae," *Astrophysical Journal* 105 (1947): 255–57; Joao Lin Yun and Dan P. Clemens, "Star Formation in Small Globules—Bart Bok was Correct," *Astrophysical Journal* 365 (December 20, 1990): L73–76;

Martin Harwit, "Neugebauer, Martz, & Leighton's Observations of Extremely Cool Stars," *Astrophysical Journal*, centennial issue 525C (1999): 1063–64; and Ken Croswell, *The Lives of Stars* (Honesdale, PA: Boyds Mill, 2009).

The Dusty Missing Link

For information on the cosmos' first dust, I drew on Eli Dwek and Isabelle Cherchneff, "The Origin of Dust in the Early Universe: Probing the Star Formation History of Galaxies by Their Dust Content," *Astrophysical Journal* 727 (2011); Anja C. Andersen, "In the Beginning: The Origin of Dust," in *Cool Stars, Stellar Systems and the Sun: Proceedings of the 15th Cambridge Workshop on Cool Stars, Stellar Systems and the Sun, AIP Conference Proceedings* 1094 (2009): 13–22; and G. C. Sloan et al., "Dust Formation in a Galaxy with Primitive Abundances," *Science* 323 (January 16, 2009): 353–55.

For examples of dust production by various types of stars and cosmic dust's interstellar journey, I used Robert Fesen et al., "The Expansion Asymmetry and Age of the *Cassiopeia A* Supernova Remnant," *Astrophysical Journal* 645, no. 1 (July 2006): 283–92; "10,000 Earths' Worth of Fresh Dust Found Near Star Explosion," "Spitzer," NASA, December 20, 2007, http://www.nasa.gov/mission_pages/spitzer/news/spitzer-20071220.html (accessed December 19, 2011); and P. Kervella et al., "The Close Circumstellar Environment of Betelgeuse. II. Diffraction-Limited Spectro-Imaging from 7.76 to 19.50 μm with VLT/VISIR," *Astronomy & Astrophysics* 531 (July 2011): A117–27.

For information on astromineralogy, I drew on the Online Etymology Dictionary's entry for *mineral*: http://www.etymonline.com/index.php?term=mineral (accessed August 2, 2011); Ernest H. Nickel, "The Definition of a Mineral," *Canadian Mineralogist* 33, no. 3 (June 1995): 689–90; Robert Hazen, "Evolution of Minerals," *Scientific American* (March 2010): 58–65; "Spitzer Sees Crystal 'Rain' in Outer Clouds of Infant Star" (May 26, 2011), "Spitzer," NASA, http://www.nasa.gov/mission_pages/spitzer/news/spitzer20110526.html (accessed December 19, 2011); and Anja C. Andersen, "Spectral Features of Presolar Diamonds in the Laboratory and in Carbon Star Atmospheres," *Astronomy & Astrophysics* 330 (1998): 1080–90.

Please see http://www.spitzer.caltech.edu/images/1949-ssc2008-15a-Spitzer-Reveals-Stellar-Family-Tree-, for Michael Werner's Continents of Creation image by Spitzer.

Sources of Direct Quotations by Page

(page 135) "*We now know that . . .*" Kaler, *Cosmic Clouds*, p. 242.

(Page 138) "*a shining fluid of a nature . . .*" Ibid., p. 11.

(page 138) "*whatever it might be.*" Sanford, "On Some Relations of the Spiral Nebulae," p. 90.

CHAPTER 6. THE COSMOS GOES GREEN

Sources by Sections

Tuning In to Molecules and Radio Whispers from the Universe

The key reference on the history of radio astronomy and its role in the emergence of astrochemistry is by pioneering radio astronomer Gerrit L. Verschuur, in his *The Invisible Universe: The Story of Radio Astronomy*, 2nd ed. (New York: Springer, 2007). The US National Radio Astronomy Observatory also has an excellent website on the history of radio telescopes: http://www.nrao.edu/index.php/learn/radioastronomy/radioastronomyhistory.

The discovery of the first interstellar molecules is recounted in Gerhard Herzberg, "Historical Remarks on the Discovery of Interstellar Molecules," *Journal of the Royal Astronomical Society of Canada* 82, no. 3 (1988): 115–27; and P. Swings, "The Pioneering Investigations in the Field of the Interstellar Molecules, 1935–1942," *Astrophysics and Space Science* 55 (1978): 263–65.

Cosmic Water Man

Townes tells his own story in Charles Townes, *How the Laser Happened: Adventures of a Scientist* (New York: Oxford University Press, 1999). Townes's key discovery papers are A. C. Cheung et al., "Detection of NH3 Molecules in the Interstellar Medium by Their Microwave Emissions," *Physical Review Letters* 21, no. 25 (December 16, 1968): 1701–1705; and A. C. Cheung et al., "Detection of Water in Interstellar Regions by Its Microwave Radiation," *Nature* 221 (February 15, 1969): 626–28. An overview of Townes's pioneering astrochemistry work is found in D. M. Rank, C. H. Townes, and W. J. Welch, "Interstellar Molecules and Dense Clouds," *Science* 174 (December 10, 1971): 1083.

For information on water masers, see Violette Impellizzeri et al., "A Gravitationally Lensed Water Maser in the Early Universe," *Nature* 456 (December 18, 2008): 927–29.

The Cosmic Sea

Two authoritative overview texts on cosmic water are Thérèse Encrenaz, *Searching for Water in the Universe* (Berlin: Springer-Praxis, 2007); and Thérèse Encrenaz, "Water in the Solar System," *Annual Review of Astronomy and Astrophysics* 46 (2008): 57–87.

An excellent summary of the view of cosmic water provided in the infrared is A. Salama and M. Kessler, "ISO and Cosmic Water," *ESA Bulletin* 104 (November 1, 2000): 30–38. Key references on the most recent discoveries in cosmic water, including in our Solar System, include "Astronomers Find Largest, Most Distant Water Reservoir," NASA, Jet Propulsion Lab press release, July 22, 2011, http://www.jpl.nasa.gov/news/news.cfm?release=2011-223 (accessed September 1, 2011); Kenneth Chang, "Evidence of Water beneath Moon's Stony Face," *New York Times*, May 26, 2011, http://www .nytimes.com/2011/05/27/science/space/27moon.html; and Kenneth Chang, "Red Planet May Be Better Known as the Wet One," *New York Times*, September 9, 2009.

Brad Scriber, "Water's Out There," *National Geographic, Water: A Special Issue*, April 2010; "Recipe for Water: Just Add Starlight," European Space Agency, September 2, 2010, http://www.esa.int/esaSC/SEMW76EODDG_index_0.html (accessed December 11, 2011); "Dusty Experiments Are Solving Interstellar Water Mystery," Royal Astronomical Society press release (April 12, 2010); and F. Dulieu et al., "Experimental Evidence for Water Formation on Interstellar Dust Grains by Hydrogen and Oxygen Atoms," *Astronomy & Astrophysics* (March 2009): A30–35.

Joining Heaven and Earth and Red Giants and White Dwarfs

A good overview of the complexity of interstellar chemistry is Eric Herbst and Ewine F. van Dishoeck, "Complex Organic Interstellar Molecules," *Annual Review of Astronomy and Astrophysics* 47 (2009): 427–80.

Several of Ziurys's articles that provide an overview of her work include Lucy M. Ziurys, "Molecular Spectroscopy of Transient Species as Probe of Interstellar Chemistry" (PhD thesis, University of California–Berkeley, 1984); Lucy M. Ziurys, "The Chemistry in Circumstellar Envelopes of Evolved Stars: Following the Origin of the Elements to the Origin of Life," *Proceedings of the National Academy of Sciences* 103, no. 33 (August 15, 2006): 12274–79; and Lucy M. Ziurys et al., "Chemical Complexity in the Winds of Oxygen-Rich Supergiant Star VY Canis Majoris," *Nature* 447 (June 2007): 1094–97.

The critical role of molecular clouds in star birth is described in a classic summary by Leo Blitz, "Giant Molecular Cloud Complexes in the Galaxy," *Scientific American* 246, no. 4 (April 1982): 84–94. A key paper on the origins of molecules from dying stars is Sun Kwok, "The Synthesis of Organic and Inorganic Compounds in Evolved Stars," *Nature* 430 (August 26, 2004): 985–91. The discovery of the first interstellar sugar is recounted in Jan Hollis et al., "Interstellar Glycolaldehyde: The First Sugar," *Astrophysical Journal* 540 (September 10, 2000): L107–10.

The presence and role of polycyclic aromatic hydrocarbons (PAHs) are summarized in C. Joblin and A. G. Tielens, eds., "PAHs and the Universe," *EAS Publications Series* 46 (2011), particularly the following chapters: A. G. Tielens, "25 Years of PAH Hypothesis," pp. 4–7, and L. J. Allamandola, "PAHs and Astrobiology," p. 306. See also PAHs researcher Lou Allamandola's NASA-Ames website https://astrobiology-beta.arc.nasa.gov/directory/people/allamandola-louis/ (accessed March 19, 2012).

For an overview of diffuse interstellar bands (DIBs), see Peter J. Sarre, "The Diffuse Interstellar Bands: A Major Problem in Astronomical Spectroscopy," *Journal of Molecular Spectroscopy* 238 (2006): 1–10.

Sources of Direct Quotations by Page

(page 159) "*The surface of the Earth . . .*" Carl Sagan, *Cosmos* (New York: Random House, 1980), p. 5.

(page 162) "*Our observations suggest . . .*" Jan Hollis, e-mail message to author, June 20, 2008. See also: Lara Clemence and Jarrett Cohen, "Space Sugar's a Sweet Find," NASA, February 7, 2005. http://www.nasa.gov/vision/universe/stargalaxies/interstellar_sugar.html (accessed June 19, 2012).

(page 163) "*could not dream up . . .*" Verschuur, *Invisible Universe*, p. 14.

(page 169) "*Charlie, how could you . . .*" Townes, *How the Laser Happened*, p. 156.

(page 191) "*These observations have . . .*" Rank, Townes, and Welch, "Interstellar Molecules and Dense Clouds," p. 1083.

PART 3: THE LIVING COSMOS

CHAPTER 7: CATCHING STARDUST

Sources by Sections

The Space-Rock Education of Scott Sandford

In addition to my interview with Dr. Sandford, the following sources informed this section: S. G. Love and D. E. Brownlee, "A Direct Measurement of the Terrestrial Mass Accretion Rate of Cosmic Dust," *Science* 262 (October 22, 1993): 550–53; Matthieu Gounelle, "The Asteroid-Comet Continuum: In Search of Lost Primitivity," *Elements* 7 (September 2011): 29–34; and Vince Stricherz, "Like a Rock: New Mineral Named for UW Astronomer," University of Washington press release, June 26, 2008, http://www.washington.edu/news/archive/id/42636 (accessed January 16, 2012).

The Birth of the Earth

Overview information on meteorites and on the history of meteoritics is from Edward Scott, "Meteorites: An Overview," *Elements* 7 (2011): 47–48; and Christopher Cokinos, *The Fallen Sky: An Intimate History of Shooting Stars* (New York: Penguin, 2010).

A very helpful, accessible overview of the way meteorites have come to help us understand the origins of the Solar System is Dante S. Lauretta, "A Cosmochemical View of the Solar System," *Elements* 7 (September 2011): 11–16.

The story of the Allende meteorite's arrival and initial study is compiled from Elbert King, *Moon Trip: A Personal Account of the Apollo Program and Its Science* (Houston: University of Houston, 1989); Arch Reid, "Elbert Aubrey King, Jr. (1935–1998)," *Meteoritics & Planetary Science* 34 (1999): 677; "Blue-White Fireball Sighted in Mexico," *Washington Post*, UPI, February 9, 1969, http://www.mnh.si.edu/onehundredyears/featured_objects/AllendeMeteorite.html; and from the Smithsonian Institution, where a large piece of Allende is on display: "Allende Meteorite," Smithsonian National Museum of Natural History, http://www.mnh.si.edu/onehundredyears/featured_objects/AllendeMeteorite.html (accessed January 17, 2011).

The scientific papers documenting Allende's age are Jamie Gilmour, "The Solar System's First Clocks," *Science* 297 (September 6, 2002): 1658–59; Yuri Amelin et al., "Lead Isotopic Ages of Chondrules and Calcium-Aluminum-Rich Inclusions," *Science* 297 (September 6, 2002): 1678–83; Hubert Reeves, "Cosmochronology after Allende," *Astrophysical Journal* 231 (July 1, 1979): 229–35; and Typhoon Lee et al., "Aluminum-26 in the Early Solar System: Fossil or Fuel?" *Astrophysical Journal* 211 (1977): 107–10.

The Men Who First Held Stardust

This story of the discovery of the first pre-solar diamonds is compiled from a wide range of perspectives.

Information about Edward Anders comes from Ursula Marvin, "Oral Histories in Meteoritics and Planetary Science: I. Edward Anders," *Meteoritics & Planetary Science* 36 (2001): A255–67; and Edward Anders, "Origin, Age, and Composition of Meteorites," *Space Science Reviews*, 3. no. 11 (March 12, 1979): 321, http://garfield.library.upenn.edu/classics1979/A1979HZ32600001.pdf. Anders's powerful memoir recounting the experience of Latvian Jews during the Holocaust is widely available from online bookstores: Edward Anders, *Amidst Latvians during the Holocaust* (Riga: Occupation Museum Association of Latvia, 2010); and I'm indebted to Dr. Anders for sharing part of an unpublished autobiography for my background information: "Chicago: 1964–1991," chap. 13, pp. 218–89. Anders's key early paper outlining the origin of meteorites in the asteroid belt between Jupiter and Mars is Edward Anders, "Origin, Age, and Composition of Meteorites," *Space Science Reviews* 3 (1964): 583–714.

The description of Solar System theorist Viktor Safronov's work is based on Alan Boss, *Looking for Earths: The Race to Find New Solar Systems* (New York: John Wiley & Sons, 1998), pp. 39–42.

The process of how pre-solar diamonds were discovered is compiled from the original research paper of Roy S. Lewis et al., "Interstellar Diamonds in Meteorites," *Nature* 326 (March 12, 1987): 160–62; Anders's description of the process is found in Edward Anders and Ernst Zinner, "Interstellar Grains in Primitive Meteorites: Diamond, Silicon Carbide, and Graphite," *Meteoritics* 28 (1993): 490–514; and two media accounts come from Patrick Huyghe, "Stardust Memories," *Discover*, November 1991, pp. 58–64; and Faye Flam, "Seeing Stars in a Handful of Dust," *Science* 253 (July 26, 1991): 380–81.

Stardust Memories

This section is based on Ann Nguyen and Scott Messenger, "Presolar History Recorded in Extraterrestrial Materials," *Elements* 7, no. 1 (February 2011): 17–22; Andrew M. Davis, "Stardust in Meteorites," *Proceedings of the National Academy of Sciences* 108 (November 29, 2011): 19142–46; Donald Clayton and Larry Nittler, "Astrophysics with Presolar Dust," *Annual Review of Astronomy and Astrophysics* 42 (2004): 39–78; and Douglas Rumble III, "Stable Isotope Cosmochemistry and the Evolution of Planetary Systems," *Elements* 7 (February 2011): 23–28.

The description of Solar System formation from stardust clumping is based on Daniel Apai and Dante S. Lauretta, "Planet Formation and Protoplanetary Dust," pp. 9–15, in *Protoplanetary Dust: Astrophysical and Cosmochemical Perspectives*, ed. Daniel Apai and Dante S. Lauretta (Cambridge: Cambridge University Press, 2010); Jonathan Williams and Lucas Cieza, "Protoplanetary Disks and Their Evolution," *Annual Review of Astronomy and Astrophysics* 49 (2011): 67–117; C. H. Chen et al., "Spitzer IRS Spectroscopy of IRAS-Discovered Debris Disks," *Astrophysical Journal Supplement Series* 166 (September 2006): 351–77; and Thorsten Kleine and John F. Rudge, "Chronometry of Meteorites and the Formation of the Earth and Moon," *Elements* 7 (February 2011): 41–46.

Sagan's Dream

The description of Sandford's cold cosmic cloud–chemistry research is based on a tour he gave me of the NASA-Ames astrochemistry lab. Related scientific papers include Michel Nuevo et al., "Formation of Uracil from the Ultraviolet Photo-Irradiation of Pyrimidine in Pure H2O Ices," *Astrobiology* 9 (2009): 683–95; Max P. Bernstein et al., "Racemic Amino Acids from the Ultraviolet Photolysis of Interstellar Ice Analogues," *Nature* 416 (March 28, 2002): 401–403; and Max Bernstein et al., "Organic Compounds Produced by Photolysis of Realistic Interstellar and Cometary Ice Analogs Containing Methanol," *Astrophysical Journal* 454 (November 20, 1995): 327–44.

Sagan's goal of taking the Miller-Urey experiment cosmic is described in James Strick, "Creating a Cosmic Discipline: The Crystallization and Consolidation of Exobiology, 1957–1973," *Journal of the History of Biology* 37 (2004): 131–80; see p. 135.

DNA from Space

This section draws on the following research papers: Zita Martins, "Organic Chemistry of Carbonaceous Meteorites," *Elements* 7, no. 1 (February 2011): 35–40; Sandra Pizzarello, "The Cosmochemical Record of Carbonaceous Meteorites: An Evolutionary Story," *Journal of the Mexican Chemical Society* 53, no. 4 (2009): 253–60; Michael P. Callahan et al., "Carbonaceous Meteorites Contain a Wide Range of Extraterrestrial Nucleobases," *Proceedings of the National Academy of Sciences* (published online August 11, 2011): 13995–98, http://www.pnas.org/content/early/2011/08/10/1106493108 (accessed August 11, 2011); Philippe Schmitt-Kopplin et al., "High Molecular Diversity of Extraterrestrial Organic Matter in Murchison Meteorite Revealed 40 Years after Its Fall," *Proceedings of the National Academy of Sciences* 107 (February 16, 2010): 2763–68; and George Cooper et al., "Carbonaceous Meteorites as a Source of Sugar-Related Organic

Compounds for the Early Earth," *Nature* 414 (December 27, 2001): 879–83.

For the organic content of comets, see Scott Sandford, "Terrestrial Analysis of the Organic Component of Comet Dust," *Annual Review of Analytical Chemistry* 1 (2008): 549–78; Scott Sandford et al., "Organics Captured from Comet 81P/Wild 2 by the Stardust Spacecraft," *Science* 314 (December 15, 2006): 1720–24; Michael Mumma and Steven Charnley, "The Chemical Composition of Comets—Emerging Taxonomies and Natal Heritage," *Annual Review of Astronomy and Astrophysics* 49 (2011): 471–524; and Matthieu Gounelle, "The Asteroid-Comet Continuum: In Search of Lost Primitivity," *Elements* 7 (2011): 29–34.

An excellent account of the arrival and collection of the Tagish Lake meteorite is James Scott Berdahl, "Morning Light—The Secret History of the Tagish Lake Fireball," (masters of science writing thesis, Massachusetts Institute of Technology, 2010).

From Eternity to Here

The original science paper outlining the possible cosmic delivery of Earth's organic material is Christopher Chyba and Carl Sagan, "Endogenous Production, Exogenous Delivery and Impact-Shock Synthesis of Organic Molecules: An Inventory for the Origins of Life," *Science* 355 (January 9, 1992): 125–31. More recent evidence is recounted in Stephen Mojzsis, "Bio-Essential Element Inventories to Earth (and Earth-Like Worlds) in Late Heavy Bombardments," paper presented at the Delivery of Volatiles & Organics: From Earth to ExoEarths in the Era of JWST, Space Telescope Science Institute, Baltimore, September 13–15, 2010, http://www.stsci.edu/institute/conference/volatile/talksList.

Information on cosmic delivery of Earth's water is from Karen Meech, "Origins of Earth's Water," presented at the Delivery of Volatiles & Organics: From Earth to ExoEarths in the Era of JWST, Space Telescope Science Institute, Baltimore, September 13–15, 2010;

and Karen Meech, "Origins of Earth's Water," STScI Webcasting, broadcast December 2, 2011, https://webcast.stsci.edu/webcast/detail.xhtml;jsessionid=FA8292C5A45296C351F4B1FE55CDECA7?talkid=2825&parent=1 (accessed April 27, 2012).

Stephen Mojzsis and T. Mark Harrison, "Vestiges of a Beginning: Clues to the Emergent Biosphere Recorded in the Oldest Known Sedimentary Rocks," *GSA Today* 10, no. 4 (April 2000), http://www.geosociety.org/pubs/gsatoday/archive/sci0004.htm (accessed January 23, 2012).

Tracing Our Cosmic Carbon Ancestry

An overview paper is Eric Herbst and Ewine F. van Dishoeck, "Complex Organic Interstellar Molecules," *Annual Review of Astronomy and Astrophysics* 47 (2009): 427–80.

Key scientific papers on the origins of the organic molecules in carbonaceous chondrites are George D. Cody et al., "Establishing a Molecular Relationship between Chondritic and Cometary Organic Solids," *Proceedings of the National Academy of Sciences* 108 (November 29, 2011): 19171–76; Keiko Nakamura-Messenger et al., "Organic Globules in the Tagish Lake Meteorite: Remnants of the Protosolar Disk," *Science* 314 (December 1, 2006): 1439–42; and Christopher Herd et al., "Origin and Evolution of Prebiotic Organic Matter as Inferred from the Tagish Lake Meteorite," *Science* 332 (June 10, 2011): 1304–1307.

For the possible stellar and interstellar origins of organic molecules that arrive on Earth, see Lori Stiles, "Arizona Radio Observatory Team Discovers Supergiant Star Spews Molecules Needed for Life," UA News, University of Arizona, July 3, 2007, http://uanews.org/node/13463 (accessed January 23, 2012).

"Astrobiologists Discover 'Sweet Spots' for the Formation of Complex Organic Molecules in the Galaxy," Rensselaer Polytechnic Institute press release, November 2, 2011, http://news.rpi.edu/update.do?artcenterkey=2941 (accessed April 27, 2012).

The discussion of the possible existence of pre-solar organic stardust is from Sun Kwok, "The Synthesis of Organic and Inorganic Compounds in Evolved Stars," *Nature* 430 (August 26, 2004): 985–91; and Sun Kwok and Yong Zhang, "Mixed Aromatic-Aliphatic Organic Nanoparticles as Carriers of Unidentified Infrared Emission Features," *Nature* 479 (November 3, 2011): 80–83.

Sources of Direct Quotations by Page

(page 200) "*was so bright we had to shield . . .*" "Blue-White Fireball Sighted in Mexico," p. 3.

(page 200) "*The people, especially the people . . .*" Ibid.

(page 212) "*They would barely make . . .*" Flam, "Seeing Stars in a Handful of Dust," p. 381.

(pages 212–13) "*no one had seriously . . .*" Lewis et al., "Interstellar Diamonds in Meteorites," p. 161.

(page 214) "*As a young graduate . . .*" Marvin, "Oral Histories in Meteoritics and Planetary Science," p. A262.

(page 214) "*Until recently people could study . . .*" Huyghe, "Stardust Memories," p. 58.

CHAPTER 8. OTHER WORLDS

This chapter draws heavily on my interviews over the past several years with the principal scientists in this story, as listed in the interviews section.

Two of the best historical overviews of the search for exoplanets were written by leading exoplanet theorist Alan Boss, who both participated in and watched the events unfold and who tells the story in a detailed chronological account in Alan Boss, *Looking for Earths: The Race to Find New Solar Systems* (New York: John Wiley & Sons, 1998); and Alan Boss, *The Crowded Universe: The Search for Living Planets* (New York: Basic Books, 2009). An accessible introduction to the broad topic of exoplanets is my book: Jacob Berkowitz, *Out of*

This World: The Amazing Search for an Alien Earth (Toronto: Kids Can Books, 2009). A concise, engaging recent overview of our understanding of exoplanets is provided by Geoff Marcy, the astronomer who's found more exoplanets than any other, in Geoffrey W. Marcy, "Other Earths and the Search for Life in the Universe," *Proceedings of the American Philosophical Society* 154, no. 4 (December 2010): 422–38.

The classic nineteenth-century perspective on the abundance of planets around other stars is Richard Proctor, *Other Worlds Than Ours: The Plurality of Worlds Studied under the Light of Recent Scientific Researches* (Akron, OH: Werner, 1870). Two key twentieth-century scientific papers that influenced astronomers' perspectives that such planets were there for the finding are Otto Struve, "The Cosmological Significance of Stellar Rotation," *Popular Astronomy* 525 (May 1945); and Gerard Kuiper, "On the Origin of the Solar System," *Proceedings of the National Academy of Sciences* 37, no. 1 (January 15, 1951).

Sources by Sections

A New Vision

The section profiling the pioneering modern exoplanet search from the mid-1970s to the late 1980s led by Bruce Campbell and Gordon Walker draws from my feature newspaper article: Jacob Berkowitz, "Lost World: How Canada Missed Its Moment of Glory," *Globe and Mail*, September 26, 2009. Walker has also described this period in a popular account: Gordon Walker, "In Search of Other Worlds," *Cosmos Magazine*, August 16, 2007; and in a scientific recollection: Gordon Walker, "The First High-Precision Radial Velocity Search for Extra-Solar Planets," December 16, 2008, Cornell University Library, http://arxiv.org/abs/0812.3169, also published in *New Astronomy Review* 56 (2012): 9–15. The other key scientific papers covering this period include Bruce Campbell, "Precision Radial Velocities,"

Publication of the Astronomical Society of the Pacific 95 (September 1983): 577–84; Bruce Campbell, G. A. H. Walker, and S. Yang, "A Search for Substellar Companions to Solar-Type Stars," *Astrophysical Journal* 331 (August 15, 1988): 902; Gordon Walker et al., "γ Cephei: Rotation or Planetary Companion?" *Astrophysical Journal* 396 (September 10, 1992): L91–94; and Alan Boss, "Proximity of Jupiter-Like Planets to Low-Mass Stars," *Science* 267, no. 5196 (June 1995): 360–62.

A fascinating perspective on the era of exoplanet errors comes from two former American Astronomical Society public relations officers who watched the astronomers squirm at press conferences: Laurence Marschall and Stephen Maran, *Pluto Confidential: An Insider Account of the Ongoing Battles over the Status of Pluto*, chap. 10 (Dallas: BenBella Books, 2009), pp. 164–81.

Dr. Seuss's Universe

In such a rapidly developing field, what was earth-shattering yesterday is often old hat a week later. For the latest exoplanet results, a good place to start is with two up-to-the-day online scientific resources. The Extrasolar Planets Encyclopedia lists all confirmed exoplanets, including links to related literature for each exoplanet: http://www.exoplanet.eu/. The California Planet Survey provides an up-to-date, comprehensive, searchable listing with a bounty of additional resources: http://exoplanets.org/cps.html.

The following are key scientific papers documenting the golden age of results from the Kepler Space Telescope and the European Southern Observatory's HARPs spectrograph revealing the abundance and diversity of exoplanets, including "super-Earths" and multi-planet solar systems as profiled in the section "Dr. Seuss's Universe": C. Lovis et al., "The HARPS Search for Southern Extra-Solar Planets. XXVIII. Up to Seven Planets Orbiting HD 10180: Probing the Architecture of Low-Mass Planetary Systems," *Astronomy & Astrophysics* 528 (2011): 112L; Jack Lissauer et al., "A

Closely Packed System of Low-Mass, Low-Density Planets Transiting Kepler-11," *Nature* 470 (February 3, 2011): 53; A. Howard, "The Occurrence and Mass Distribution of Close-In Super-Earths, Neptunes, and Jupiters," *Science* 330 (October 29, 2010): 654; "Fifty New Exoplanets Discovered by HARPS: Richest Haul of Planets So Far Includes 16 New Super-Earths," European Space Agency press release eso1134, European Southern Observatory, September 12, 2011, http://www.eso.org/public/news/eso1134/ (accessed December 11, 2011); Joachim Wambsganss, "Astronomy: Bound and Unbound Planets Abound," *Nature* 473 (May 19, 2011): 289–91; Hagai B. Perets and M. B. N. Kouwenhoven, "On the Origin of Planets at Very Wide Orbits from the Re-Capture of Free Floating Planets," *Astrophysical Journal* 750 (April 2012): 83; Laurance R. Doyle et al., "Kepler-16: A Transiting Circumbinary Planet," *Science* 333 (September 16, 2011): 1602; Kevin J. Walsh et al., "A Low Mass for Mars from Jupiter's Early Gas-Driven Migration," *Nature* 475 (July 14, 2011): 206–209; and David Nesvorný, "Young Solar System's Fifth Giant Planet?" *Astrophysical Journal Letters* (September 13, 2011). The Hubble Deep Field image can be viewed here: http://www.hubblesite.org/newscenter/archive/releases/1996/01/.

Alien Earth

Veteran science writer Michael Lemonick wrote the *Time* magazine cover story about the discovery of the first exoplanet in 1995, and he tells the history of the search for an alien Earth in *Mirror Earth: The Search for Our Planet's Twin* (New York: Bloomsbury, 2012).

For the detailed backstory on the origins of the Kepler Space Telescope, the following journal articles mark seminal steps in the process: Frank Rosenblatt, "A Two-Color Photometric Method for Detection of Extra-Solar Planetary Systems," *Icarus* 14 (1971): 71–93; W. J. Borucki and A. L. Summers, "The Photometric Method of Detecting Other Planetary Systems," *Icarus* 58 (1984): 121; James Kasting, Daniel Whitmire, and Ray Reynolds, "Habitable Zones around Main Sequence Stars," *Icarus* 101 (1993): 108; David Koch et al., "System Design of a Mission to Detect Earth-Sized Planets in the Inner Orbits of Solar-Like Stars," *Journal of Geophysical Research* 101, no. E4 (April 25, 1996); and Ron Cowen, "Jumpy Stars Slow Hunt for Other Earths," *Nature* 477 (September 6, 2011): 142–43, http://www.nature.com/news/2011/110906/full/477142a .html. Finally, one of the many press releases claiming the discovery of the first habitable-zone Earth-like exoplanet is "First Habitable-Zone Super-Earth Discovered in Orbit around a Sun-Like Star," Carnegie Institution for Science, December 5, 2011, http://carnegiescience .edu/news/first_habitablezone_superearth_discovered_orbit_around_ sunlike_star.

Sources of Direct Quotations by Page

(page 241) "*The province of the student* . . ." George Ellery Hale, *The Study of Stellar Evolution: An Account of Some Recent Methods of Astrophysical Research* (Chicago: University of Chicago Press, 1908), p. 4.

(page 244) "*utterly unique* . . ." Peter Ward and Donald Brownlee, *Rare Earth: Why Complex Life Is Uncommon in the Universe* (New York: Springer-Verlag, 2000) p. xxiv.

(page 250) "*It is quite hard* . . ." Walker, "First High-Precision Radial Velocity Search for Extra-Solar Planets," p. 2.

(page 252) "*I probably won't* . . ." Marschall and Maran, *Pluto Confidential*, pp. 170–71.

CHAPTER 9. DARWIN AND THE COSMOS

Many years ago, I read and was inspired by a book that led me to thinking about the issues discussed in this chapter: Brian Swimme, *The Universe Is a Green Dragon: A Cosmic Creation Story* (Santa Fe, NM: Bear, 1985). The history of the concept of cosmic evolution is covered in Steven J. Dick, "Cosmic Evolution: History, Culture, and Human Destiny," in *Cosmos & Culture: Cultural Evolution in a Cosmic Context*, ed. Steven J. Dick and Mark Lupisella (Washington, DC: National Aeronautics and Space Administration, Office of External Relations, History Division, 2009). For the reference to Darwin's "gone cycling on," I'm indebted to Timothy Ferris, who noted Darwin's use of the term in his book *The Whole Shebang: A State-of-the-Universe(s) Report* (New York: Simon & Schuster, 1997), p. 175.

Sources by Sections

The Biological Big Bang

The section on Alexander Dalgarno is based on my interview with him and on the articles: Alexander Dalgarno, "A Serendipitous Journey," *Annual Review of Astronomy and Astrophysics* 46 (2008): 1–20; Alexander Dalgarno, "The Growth of Molecular Complexity in the Universe," *Faraday Discussions* 133 (2006): 9–25; and Volker Bromm and Richard B. Larson, "The First Stars," *Annual Review of Astronomy and Astrophysics* 42 (2004): 79–118. Current research goals on the eras of cosmic dawn and the cosmic dark ages are outlined in National Research Council of the National Academies, *New Worlds, New Horizons in Astronomy and Astrophysics* (Washington, DC: National Academies Press, 2010).

What Is "Life"?

In writing this section, I drew on Steven Benner, *Life, the Universe, and the Scientific Method* (Gainesville, FL: Ffame Press, 2009); Steven Benner "Defining Life," *Astrobiology* 10 (2010): 1021–30; Stephane Tirard, Michel Morange, and Antonio Lazcano, "The Definition of Life: A Brief History of an Elusive Scientific Endeavor," *Astrobiology* 10 (2010): 1003–1009; C. E. Cleland and C. F. Chyba, "Defining 'Life,'" *Origins of Life and Evolution of Biospheres* 35 (2002):

333–43; and Antonio Lazcano, "Which Way to Life?" *Origins of Life and Evolution of Biospheres* 40 (2010): 161–67.

Alonso Ricardo and Steven Benner, "The Origin of Proteins and Nucleic Acids," in *Planets and Life: The Emerging Science of Astrobiology*, ed. Woodruff T. Sullivan III and John Baross (Cambridge: Cambridge University Press, 2007) p. 154.

For a fascinating look at efforts to trace back our molecular heritage, see Chiaolong Hsiao et al., "Peeling the Onion: Ribosomes Are Ancient Molecular Fossils," *Molecular Biology and Evolution* 26 (2009): 2415–45; and the Center for Ribosomal Origins and Evolution, where a large research team is working to rewind the tape of life at the molecular level: http://astrobiology.gatech.edu/home (accessed April 30, 2012).

For a full discussion of the history of the development of the RNA world theory and the current debate, see Benner, *Life, the Universe, and the Scientific Method*, chaps. 4 and 5.

Life as a Cosmic Continuum

This section draws on George Wald, "The Origins of Life," *Proceedings of the National Academy of Sciences* 52, no. 2 (August 1964): 595–611; Chris P. McKay, "What Is Life—and How Do We Search for It in Other Worlds?" *PLoS Biology* 2 (September 2004): 1260–63; Norman R. Pace, "The Universal Nature of Biochemistry," *Proceedings of the National Academy of Sciences* 98 (2001): 805–808; Pascale Ehrenfreund and Mark A. Sephton, "Carbon Molecules in Space: From Astrochemistry to Astrobiology," *Faraday Discussions* 133 (2006): 277–88; and National Research Council of the National Academies, *The Limits of Organic Life in Planetary Systems* (Washington, DC: National Academies Press, 2007).

Astrobiology pioneer Lynn Rothschild's "Replaying the Tape" lecture to her astrobiology and space exploration class at Stanford University can be found at https://humbio.stanford.edu/node/2427 (accessed April 30, 2012).

The study of element-based phylogenetics is the new field of *paleoecophylostoichiometrics*: *paleo*: "ancient"; *eco*: "environment"; *phylo*: "genetic relationship"; *stochiometrics*: "measurement of relative quantities"; see Aditya Chopra et al., *Palaeoecophylostoichiometrics: Searching for the Elemental Composition of the Last Universal Common Ancestor*, in Australian Space Science Conference Series: 9th Conference Proceedings. Full Refereed Proceedings DVD, National Space Society of Australia Ltd., ISBN 13: 978-0-9775740, 2010.

For thoughts on the cosmic selection process of our genetic code, see Paul G. Higgs and Ralph E. Pudritz, "A Thermodynamic Basis for Prebiotic Amino Acid Synthesis and the Nature of the First Genetic Code," *Astrobiology* 9, no. 5 (2009): 483–89; and Gayle K. Philip and Stephen J. Freeland, "Did Evolution Select a Nonrandom 'Alphabet' of Amino Acids?" *Astrobiology* 11, no. 3 (2011): 235–40.

For background on possible abiotic cell membrane precursors, see Sandra Pizzarello, "The Cosmochemical Record of Carbonaceous Meteorites: An Evolutionary Story," *Journal of the Mexican Chemical Society* 53, no. 4 (2009): 253–60; George D. Cody et al., "Establishing a Molecular Relationship between Chondritic and Cometary Organic Solids," *Proceedings of the National Academy of Sciences* 108 (November 29, 2011): 19171–76; and "Protocells," Center for Fundamental Living Technology, http://flint.sdu.dk/index.php?page=protocell.

Bunsen and Kirchhoff's Gift

For an overview of exoplanet atmospheric science, see Sara Seager and Drake Deming, "Exoplanet Atmospheres," *Annual Review of Astronomy and Astrophysics* 48 (2010): 631–72. For proposed missions to study possible living exoplanets, see C. S. Cockell et al., "Darwin—A Mission to Detect and Search for Life on Extrasolar Planets," *Astrobiology* 9, no. 1 (January–February 2009): 1–22; and see the entire issue of *Astrobiology* 10, no. 1 (January–February 2010), especially the article by M. Fridlund et al., "The Search for Worlds Like Our Own," pp. 5–19.

An excellent overview of the history and challenging future of direct exoplanet imaging is Paul Kalas, "Planetary Systems Revealed through Direct Imaging," *Hipparchos* 2 (September 2011): 23–28.

For an overview of how NASA discusses its mission in finding an alien Earth, see PlanetQuest on NASA's Jet Propulsion Laboratory's site: http://planetquest.jpl.nasa.gov/.

The description of the history of searching for living planets based on their atmospheric characteristics is based on J. E. Lovelock, "A Physical Basis for Life Detection Experiments," *Nature* 207 (August 7, 1965): 568–70; and Dian R. Hitchcock and James E. Lovelock, "Life Detection by Atmospheric Analysis," *Icarus* 7 (1967): 149–59.

A link between exoplanet geology and atmospherics is found in Nikku Madhusudhan, "A High C/O Ratio and Weak Thermal Inversion in the Atmosphere of Exoplanet WASP-12b," *Nature* 469 (January 6, 2011): 64–67. See also Michael F. Sterzik, Stefano Bagnulo, and Enric Palle, "Biosignatures as Revealed by Spectropolarimetry of Earthshine," *Nature* 483 (March 1, 2012): 64–66; Shawn D. Domagal-Goldman et al., "Using Biogenic Sulfur Gases as Remotely Detectable Biosignatures of Anoxic Planets," *Astrobiology* 11, no. 5 (2011): 419–41; and David J. Des Marais et al., "Biosignatures and Planetary Properties to Be Investigated by the TPF Mission," NASA-JPL Publication 01-008, Rev. A, October 2001 available online at http://planetquest1.jpl.nasa.gov/TPF/TPF_Biomrkr_REV_3_02.pdf.

An Ancient View with Stardust Eyes

I visited Guanajuato, Mexico, from January to June 2012. Presently, the University of Guanajuato offers nightly astronomy viewing from the roof of its main downtown campus building.

Sources of Direct Quotations by Page

(page 227) "*We have had a century . . .*" Wald, "Origins of Life," p. 595.

(page 284) "*a revolution in our . . .*" Dimitar Sasselov, "Two

Separate Quests, One to Discover Habitable Worlds, the Other to Synthesize Artificial Organisms, Now Unite to Redefine 'Life' and Place in the Universe," SeedMagazine.com, March 14, 2011, pp. 66–67, http://seedmagazine.com/content/article/on_discovering_life/ (accessed February 22, 2012).

(pages 288–89) "*This is no accident,*" Benner, *Life, the Universe, and the Scientific Method*, p. 23.

(page 290) "*Life is a self-sustaining chemical system . . .*" Ibid., p. 24.

(page 292) "*If the origin of life . . .*" Lazcano, "Which Way to Life?" p. 166.

(pages 292–93) "*Though no evidence . . .*" J. Peretó, J. Bada, and A. Lazcano, "Charles Darwin and the Origin of Life," *Origins of Life and Evolution of the Biosphere* 39 (2009): 404.

(page 293) "*The human genome . . .*" Ricardo and Benner, "Origin of Proteins and Nucleic Acids," p. 154.

(page 296) "*Different life forms . . .*" McKay, "What Is Life?" p. 1262.

(page 297) "*What drives us . . .*" Jacob Berkowitz, "The New Age in Astronomy: Ottawa Native Spots Jupiter-Sized Exoplanet, a Mere 500 Light-Years Away," *Ottawa Citizen*, September 10, 2006.

(page 304) "*the contrast between the apparent . . .*" Paul W. Merrill, "Stars as They Look and as They Are," *Publications of the Astronomical Society of the Pacific* 38, no. 221 (1926): 14.

(page 306) "*To the ancient Aztecs . . .*" Octavio Paz, *The Labyrinth of Solitude: Life and Thought in Mexico*, trans. Lysander Kemp (New York: Grove, 1961), p. 56.

INDEX